AIRCRAFT INSTRUMENTS

Prepared by

NAVAL AIR TECHNICAL TRAINING COMMAND

and the

U. S. NAVY TRAINING PUBLICATIONS CENTER

Memphis, Tennessee

Published by

BUREAU OF NAVAL PERSONNEL

NAVY TRAINING COURSES
NAVPERS 10333–A

UNITED STATES
GOVERNMENT PRINTING OFFICE
WASHINGTON : 1954

PREFACE

This book is written to present to the enlisted men of Naval Aviation essential information on procedures used in the installation, operation, and maintenance of aircraft instruments. This is one of a series of books designed to furnish them with the necessary information to perform their aviation duties.

Starting with an introduction to instruments, this book discusses pressure and force as related and used in certain instruments, electrical remote indicating instruments, thermometers, tachometers, fuel gages and flowmeters, compasses, gyroscopes, automatic pilots, navigation instruments, and it further contains a final chapter on testing and maintenance.

Instrument maintenance and trouble shooting procedures are discussed in a way which will provide personnel assigned to instrument shops information necessary for them to perform general instrument maintenance work. A knowledge of aircraft instruments is of primary importance to Aviation Electrician's Mates and, particularly, to those holding Emergency Service ratings. A detailed listing of the qualifications required for advancement of AE's is included in appendix II. These are taken from the Manual of Qualifications for Advancement in Rating, NavPers 18068 (Be vised).

As one of the Navy Training Courses, this book represents the joint endeavor of the Naval Air Technical Training Command, the U. S. Navy Training Publications Center, Memphis, Tennessee, and the Training Publications Section of the Bureau of Naval Personnel.

M882886

READING LIST

NAVY TRAINING COURSES

Aviation Electrician'* Mate, Vol. 1, NavPers 10319 Aviation Electrician's Mate, Vol. 2, NavPers 10320 Electricity, NavPers 10622-B

Advanced Work in Aircraft Electricity, NavPers 10316 Aviation Supply, NavPers 10394-B Blueprint Reading, NavPers 10077 Hand Tools, NavPers 10306-A

OTHER PUBLICATIONS

- Handbook of Aircraft Instruments, Vol. 1, NavAer 05-1-600 Handbook of Aircraft Instruments, Vol. 2, NavAer 05-1-500-A Aircraft Instruments Bulletins

For sale by the Superintendent of Documents, D. 8. Government Printing Office Washington 25, D. C. - Price $2

CONTENTS CHAPTER 1

Pan

Introduction to instruments ._. 1

CHAPTER 2

Pressure 5

CHAPTER 3

Remote reading instruments. 17

CHAPTER 4

Pressure gages 33

CHAPTER 5 .

Air pressure measuring instruments 57

CHAPTER 6

Thermometers.. 73

CHAPTER 7
Tachometers _ 89
CHAPTER 8
Fuel gages and flowmeters 101
CHAPTER 9
Compasses 117
CHAPTER 10
Gyroscopes. _ 139
CHAPTER 11
Automatic pilots ' 163
CHAPTER 12
Navigation instruments 199
CONTENTS—Continued CHAPTER 13
Page
Testing and maintenance 221
APPENDIX I
Answers to quizzes 260
APPENDIX II
Qualifications for advancement in rating 270
Index ... 279
*

AIRCRAFT INSTRUMENTS

•

CHAPTER

INTRODUCTION TO INSTRUMENTS

Back at the turn of the century, when aviation's pioneers were making their first airplane flights, they had no instruments in their flimsy craft. These hardy men were feeling their way around in a new world. There was no backlog of experience upon which they could lean. However, as soon as they acquired a degree of certainty about how airplanes should be manipulated, they immediately became instrument-conscious.

Early model airplane engines had a tendency to quit running at the most critical moments. These stoppages were most often caused by failures of fuel or lubrication systems. Hence, it was only natural that engine instruments \'7bpower plant instruments), to warn pilots of engine trouble in advance of disaster, were developed first.

As better airplanes were built and man perfected his ability to fly them, longer and longer flights were attempted. But increases in the range of airplanes inevitably brought new problems to pilots. They could no longer depend on guesswork when it came to determining how their airplanes were behaving in the air. They found themselves up against the whims and fickleness of the weather. They discovered that it was all too easy for an aviator to lose himself—even over familiar territory—when visibility was poor; and that flying over strange terrain was hazardous by day and impossible by night, even in the fairest weather. . Flight instruments and navigation instruments were developed to help pilots over these stumbling blocks. Flight instruments tell what must be known about the altitude, attitude, and performance of an airplane as a flying unit— whether it is at a safe height, climbing, diving, turning or

banking, and how fast it is moving through the air. Navigation instruments provide information about the location of the airplane with respect to the earth—or in relation to the sun, moon, and stars, so that the earth-wise position can be calculated.

PRECISION INSTRUMENTS

Present and future high-speed high-performance aircraft require much more precise instrumentation than the aircraft of a few years ago. To attain this greater precision, the mechanical methods of instrumenting aircraft attitudes, pressures, fuel and oil capacity and consumption are being replaced by electromechanical and electronic devices. Maintenance of the design accuracy of the newer aircraft instruments requires that more exacting standards be met in aircraft instrument maintenance procedures and techniques.

Instrument maintenance personnel must be well-versed in basic hydraulic, mechanical and gyroscopic principles. In addition, they must understand the electrical and electronic principles involved in aircraft instrument applications. Previous instrument maintenance techniques required only basic field testing and equipment exchange. The complexity of the new type aircraft instruments, such as the P-l and GS automatic pilots, and the precise application of these instrument systems to a particular type of aircraft do not readily permit just the replacement of a malfunctioning component of the system.

Application of such instruments to a specific aircraft requires that adjustments and test procedures peculiar to that aircraft be followed in the course of replacement of the malfunctioning components. On-the-spot repair of the defective component, whenever possible, is also desirable. The repair of the originally installed unit will, in most cases, be practicable when the replacement of a vacuum tube, a resistor, a capacitor, or a small subassembly will produce a serviceable instrument with a minimum delay in operational use of that aircraft.

The design of new types of instrument tools and test equipment has followed closely the design of new aircraft

instruments. Since the instruments themselves are complicated, the tools and test equipment necessary to provide the technician with a yardstick by which to apply his knowledge are likewise complicated. Some of these items of test equipment, almost as intricate as the instrument system itself, heretofore, would probably have been considered of a "laboratory" nature. With such systems as automatic pilots, fuel-quantity gages, fuel flowmeters, and other instruments employing the combined use of sensing elements, amplifiers and indicators, the ability to determine the required parts to be replaced in any one unit is no longer a "screw driver and pliers" maintenance procedure.

TECHNICIAN'S RESPONSIBILITY

The accuracy imposed on the instruments used in highspeed aircraft requires assurance that such instruments are giving correct indications. The necessity for preventive maintenance is exemplified by such instruments as percent type tachometers, tailpipe temperature indicators, gyro attitude and control indicators, and fuel quantity and fuel flow systems. The existence of excessive errors in such instrument systems has a direct relation to flight safety and to efficient aircraft performance. It cannot be assumed that borderline instrument errors are acceptable for the successful operation of high-speed aircraft.

Assurance of correct aircraft instrument indication can be attained only by periodic functional testing. Instrument testing requires time, and advantage must be taken of each inspection check to accomplish instrument operational and functional tests.

Thorough acquaintance with the operation and use of aircraft instruments and associated maintenance tools and test equipment is of personal concern to the aviation instrument mechanics. The responsibility for proper handling of instruments and associated tools must also be of personal concern. Careful handling will help to assure better operation and longer trouble-free service life.

The designation "delicate instruments" has always been applicable to aircraft instruments and must be given the utmost consideration. The same careful handling must be given instruments that are being returned to overhaul activities. A little extra care in providing protection for the malfunctioning instrument being returned to overhaul will prevent incurring damage in transit. Such damage could be serious enough to result in scrapping the instrument, thereby losing its future availability when a replacement is required.

\

QUIZ

L What aircraft instruments were developed first?

2. The instruments developed to tell the pilot the altitude, attitude, and performance of the aircraft are called instruments.

3* What instruments provide information about the plane's location with respect to the earth?

4. Aircraft instruments have become through the years.

a* Simpler.

b. Larger.

c. Smaller.

d. More complex.

5. Border-line Instrument errors are for the successful operation of high-speed aircraft.

6. Assurance of correct aircraft instrument indication can be attained only by

7. Most early engine failures were caused by faults in the

or systems.

8. Flight instruments do not tell about

a. Altitude.

b. Attitude.

c. Manifold pressure.

d. Plane performance.

9. will help to assure better instrument operation and

longer service life.

CHAPTER

PRESSURE FORCE VS. PRESSURE

Force is the action of one body on another tending to change the state of motion of the body acted upon. Force may tend to move a body at rest, or it may tend to increase or decrease the motion of a moving body or change its direction of motion. The application of force may result in one or more of the following stresses in the body acted upon: tension, pressure or compression, torsion, and shear.

A force can be measured in three ways: by the weight it can support, by its ability to stretch an elastic body, and by its ability to move a body. In physics, force is ordinarily measured in pounds or kilograms. Any force which changes the state of motion of a body is performing work upon that body. The amount of work accomplished is equal to the product of the force applied to the body and the distance through which it is moved, providing the displacement is in the direction of the force.

Pressure is force per unit area and is usually measured in pounds per square inch. For example, if the whole pull or push of an object is equal to 10 pounds, then you'd say the object was being acted upon by a total force of 10 pounds. But suppose the 10-pound force is pushing on one side only of the object in question, and that the push of this force is distributed evenly over the whole area of this particular side. Again, suppose the side being pushed has an area of 10 square inches. Since the force is divided evenly over this surface, each separate square inch of the side is being subjected to a force of 1 pound. Thus, 1 square inch is the unit area selected for the measurement of pressure. Since

each square inch is being pushed by one-tenth of the 10 pound force—or 1 pound—we know that the pressure is 1 pound per square inch. It just happens that the unit of measurement selected here was pounds per square inch. It may as well have been grams per square centimeter or tons per square foot.

For some purposes another system for indicating pressure is often used. This system expresses pressure in inches op mercury and will be discussed under fluid pressure. This brings us to the two types of pressure we are to discuss in relation to aircraft instruments: atmospheric pressure and fluid pressure. Instruments are operated by pressure either mechanically, electrically, by vacuum, or by gyro directly or through remote indicating devices.

ATMOSPHERIC PRESSURE

The atmosphere in which we live and breathe is the envelope surrounding the earth that extends upwards to about 200 miles above the surface of the earth. The atmosphere is divided into four parts—the troposphere, the stratosphere, the ionosphere, and the exosphere. The troposphere extends from sea level to about 8 miles up. The stratosphere extends from 8 miles to about 60 miles up. The ionosphere extends from 60 miles to about 120 miles up. The exosphere extends from 120 miles up. No upper linear limit has been established for this layer.

Air is a mixture of gases, and the weight of these gases which compose the atmosphere exerts a pressure against the surface of the earth. Thus, any area on the earth supports the weight of a column of air which extends above it for a distance of 200 miles, or the depth of the atmosphere. By direct measurement, it has been found that an area of 1 square inch at sea level supports a vertical column of air that weighs 14.7 pounds. At higher altitudes this force is less because the column of air supported is shorter. Therefore, the pressure exerted by the weight of the atmosphere on any area grows correspondingly smaller with any increase in altitude above sea level because of the lessened weight of the column

of air that rises vertically above it. Think of the atmosphere as being made up of a great pile of layers of air. The bottom layers are under the greatest pressure. But the higher you go, the fewer layers there will be above you pressing down, and the lower the pressure will be.

Another important factor affecting the atmosphere is the temperature. The sun sends out heat and light radiation. This radiation warms the air and, because it is warmed, its density is less so it will rise. In rising to a region of less pressure, air will expand. When air expands its temperature drops. Both on account of this expansion and because of the greater distance from the secondary source of heat on the earth, at higher altitudes in the troposphere the temperature is less.

For convenience, a standard atmosphere has been adopted. This assumes certain values of temperature and pressure for various altitudes. With the temperature and pressure decided upon, the density at various altitudes is determined. The standard temperature at sea level is 59° F., while the pressure in inches of mercury is 29.92.

Atmospheric pressure changes give valuable information concerning weather forecasting and also are used for other purposes in conjunction with aircraft instruments. For example, changing barometric pressure is incorporated in the readings of an altimeter. A barometer measures atmospheric pressure, and can be made to indicate changes in pressure.

MERCURIAL BAROMETER

Many years ago a scientist performed an experiment in an attempt to measure the pressure of the atmosphere. He took a long glass tube, sealed it at one end, and filled it full of mercury. Holding his thumb over the open end he tipped the tube upside down and placed it upright in a mercury-filled vessel. The column of mercury settled down somewhat, leaving a vacuum between its top surface and the sealed end of the tube as shown in figure 1. The experiment showed that atmospheric pressure at sea level would

hold up a column of mercury approximately 30 inches high under average conditions. This device, in reality a simple form of mercurial barometer, is called a torricellian tube.

The apparatus developed from the one this scientist employed came to be known as a mercurial barometer, and atmospheric pressure is generally measured in inches of mercury to this day.

Since it is atmospheric pressure that is supporting the column of mercury, what happens when atmospheric pres-

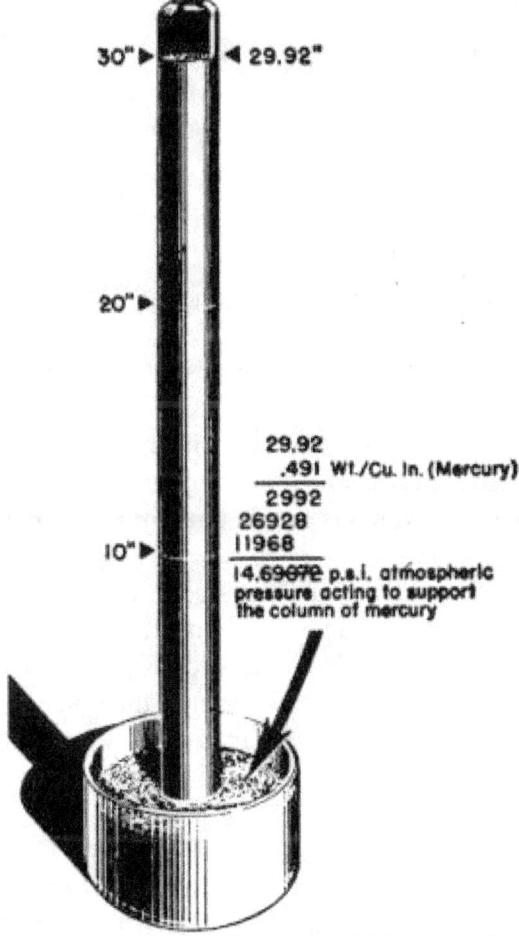

Flgure 1.—A simple mercurial baromefer. 8

sure increases or decreases? You guessed it. The length of the mercury column will increase or decrease, too, and these changes in length can easily be measured. The rest is simple. You can determine the atmospheric pressure in your vicinity in terms of the length of the column of mercury, and express it in any convenient units—depending on the kind of measuring scale used. As mentioned before, inches Hg. (inches of mercury) is the measuring system used on some aircraft instrument dials.

ANEROID

Many aircraft instruments are operated by a pressure-sensitive diaphragm capsule, or aneroid. The word "aneroid" is derived from Greek, and means "not wet." Maybe you recall the word aneroid as a term used frequently to describe a common type of barometer that uses no liquid— hence aneroid barometer.

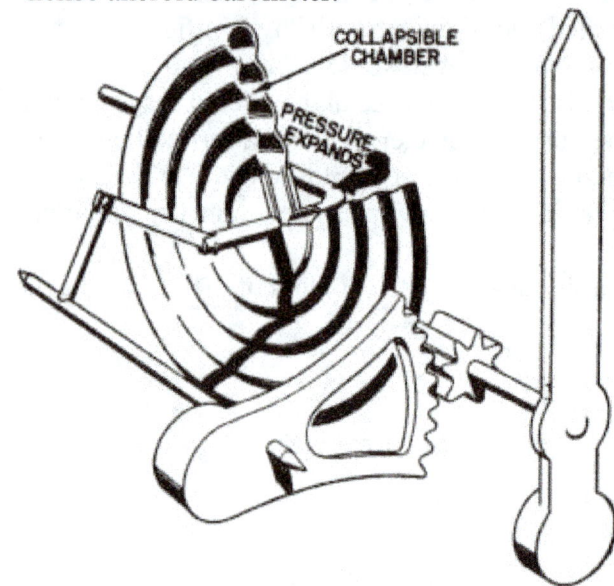

Figure 2.—Presture-senjltive aneroid.

Most aneroids in aircraft instruments look like the case of a pocket watch that has been corrugated with circular

ridges and hollows, as diagrammed in the part labeled "collapsible chamber" in figure 2. An aneroid is a sealed element and has a tendency to collapse or expand depending on the relation of the pressure sealed inside the aneroid to the existing pressure surrounding the aneroid.

Note that in figure 2 the inner side of the aneroid capsule is fixed to the frame of the instrument, and that the outer side. is attached to a post which is linked to the pointer-operating mechanism.

The casting of an aneroid pressure-measuring gage is usually sealed airtight. The pressure you wish to measure is introduced into the sealed case through a tube connection. As the pressure in the casing increases, the aneroid is compressed and its outer side is pushed inward. If the pressure in the casing is decreased, the pressure inside the aneroid pushes the outer side of the aneroid away from the inner side. In either circumstance, the post moves with the motion of the outer side of the aneroid, and causes the pointer needle to swing in the proper direction over the instrument dial.

Uses of the Aneroid

Since the aneroid barometer shows differences in pressures, and since the atmospheric pressure decreases with elevation, you can use an aneroid barometer to show your airplane's elevation. All that is necessary is to put a scale showing elevation in feet on the face of the dial instead of pressure-measuring units—and you have an altimeter.

That is not all you can do with barometers in the cockpit. One simple type of rate of climb indicator is somewhat similar to a barometer with a very slow leak in its capsule (as you see in figure 3) and with a different scale on its dial. The metal capsule has a very small hole in it so that, as the pressure outside the capsule changes, there is also a more gradual change of the pressure inside the capsule. Over a period of time these pressures will equalize. But when an airplane climbs or descends, the rate of pressure change is faster than the leak can handle immediately. Thus, your airplane is climbing, the pressure inside the capsule is—for

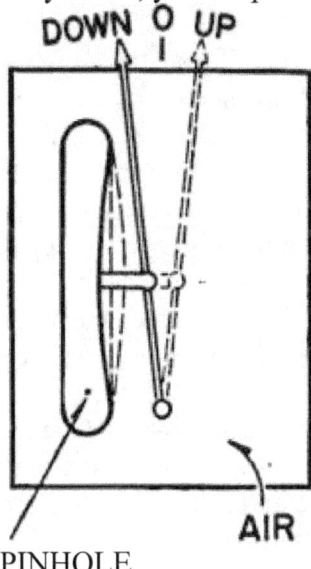

PINHOLE

Figure 3.—Diagram of a rate of climb indicator.

the time being—greater than the pressure outside. The dial indicator will then show, on its special scale, a rate of climb of so many feet per minute. When your airplane descends, the pressure inside the capsule is temporarily less than that outside it, and the indicator shows descent in feet per minute. In level flight the inside and outside pressures equalize and the instrument reads zero.

BOYLE'S LAW

A gas differs from a liquid in that a liquid is almost incompressible, while a gas is readily compressible. If temperature is kept constant, gases show a definite relationship between volume and pressure. Boyle, a British scientist, attempted to learn how the volume of a gas is affected by pressure on it. This relationship is known as Boyle's Law and states that "if the temperature remains constant, the pressure exerted by a confined gas varies inversely as the volume." In other words, the product of the pressure and the volume of a given gas remains constant.

Suppose we have a pressure in a 12-inch cylinder of 15 pounds per square inch and that we suddenly compress the gas into 6 inches of the cylinder. We do not lose any of the gas but it is occupying only one-half the volume it occupied before. Hence, the pressure is increased to 30 pounds per square inch, or just double the original pressure. If we wish to exert more force on the piston and compress the gas into 4 inches of the cylinder, then we have a pressure of 45 pounds per square inch or three times as much as it was in the beginning, and the volume has decreased to one-third the original volume.

Boyle also found that the volume of a given quantity of gas and the pressure on it are strongly affected by temperature changes. A gas will expand when heat is applied. Air or gas inside a balloon will expand as its temperature is increased. It may be stated that the pressure of a gas varies directly with its absolute temperature when the volume of a gas remains constant.

Various applications of Boyle's Law may be found in aircraft instruments, all using varying degrees of volume, pressure, or temperature.

FLUID PRESSURE

A variety of aircraft instruments are operated by fluid pressure. Fluids are substances which may be made to flow or change their shapes by the application of moderate pressures. Internal friction in fluids is very small; in some fluids, for ordinary calculations, it may be neglected. When this is done the fluid is being regarded as if it were a "perfect" fluid. Although actually there are no "perfect" fluids, the errors involved are usually small.

In fluids, at any point the pressure is the same in all directions. If a surface is immersed in a stationary fluid, the pressure of the fluid will be at right angles to the stationary surface. Some interesting things happen when a fluid is confined in a leakproof container. Under such conditions, an increase in pressure on any part of the confined fluid will cause the same increase in pressure throughout the fluid. This is known as Pascal's principle. It doesn't matter what shape the container happens to have, or what kind of a fluid is in it, as long as there are no holes or exits through which the fluid can escape so as to relieve the pressure.

Since an increase in pressure on any part of a confined fluid will produce the same pressure throughout the fluid, then the application of a small force on a small area will result in a much larger force on a large area.

This is the principle upon which a number of hydraulic mechanisms work. One everyday example you'll recognize immediately in this connection is the hydraulic brake system of a modern automobile. The tubes, pistons, and cylinders which make up such a brake system are really an enclosed container filled with brake fluid. By pushing lightly on the brake pedal you operate a small piston, thereby applying pressure on a small area of the fluid surface. The fluid immediately exerts a much greater force on the larger pistons operating the brake-bands on each wheel, and the speeding car is quickly brought to a stop.

The successful operation of modern airplanes depends in many ways on fluid pressure systems. Oil, air and fuel for supercharged, high-power aircraft engines are fed in under pressure. The de-icing equipment on wings and stabilizers works by air under pressure. Retractable landing gear, bomb-bay doors, tail wheels, and flaps are operated by hydraulic pressure in many types of military airplanes.

Accurate information about fluid pressure systems is of great importance to aviators, as you can see. If the fuel pressure is too low, an engine may conk out. Subnormal oil pressure might well mean that an engine will tear itself apart. Improper functioning of de-icers could cause a forced landing or a crackup.

You can't test an airplane's fluid pressure systems by watching how far they will squirt streams of fluid—or, at least you can't more than once. Some excellent instruments have been invented, however, to give you the lowdown on pressure-system operation. Such instruments are generally known as pressure gages, and their working principles are all somewhat similar.

Pressure gages are connected into the particular systems for which they are designed by means of small pipelines. Each gage actually becomes a part of the "fluid container" of its system, and is supplied with a small quantity of fluid at the uniform system pressure.

BOURDON TUBE

Some types of aircraft instruments are operated by a device known as the Bourdon tube. In principle, the Bourdon tube is a grownup version of an old familiar Hallowe'en novelty— the rolled-up paper tube which uncoils when you blow into one end of it and which, when you stop blowing, curls up into its original shape because of the little wire spring inside. Blowing gently makes the paper tube uncoil only part way, but a good stiff puff pushes it out full length.

The Bourdon tube in an aircraft instrument is made of metal tubing, oval or somewhat flattened in cross section,

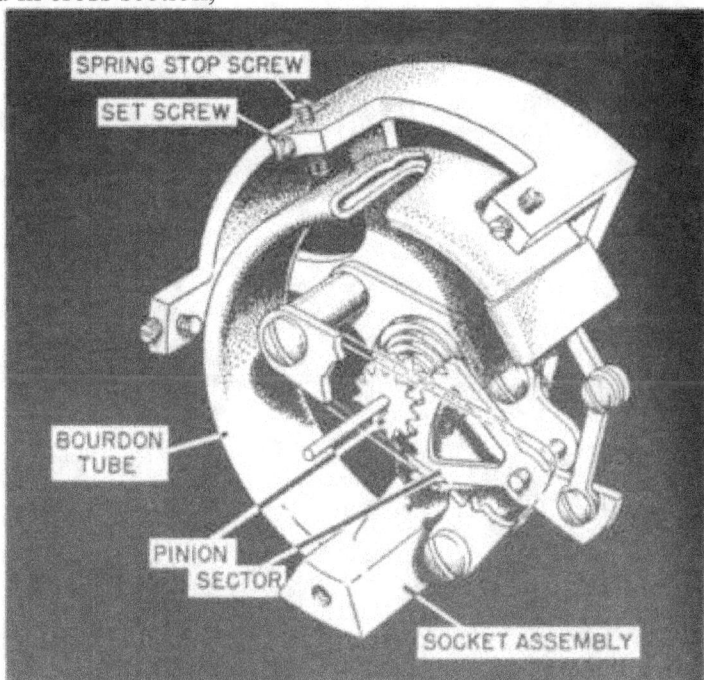

Figure 4.—Diagram of Bourdon tube pressure gage.

closed at one end and mounted rigidly in the instrument case at its other end. Look at figure 4 and notice that the Bourdon tube is curved like a crescent moon. The fluid, on which you want to get a pressure reading, is introduced into the fixed end of the Bourdon tube by a small pipe leading from the fluid system to the instrument. As with your paper Hallowe'en toy, the greater the inside fluid pressure, the more the Bourdon tube tends to straighten out. When the inside pressure drops, the natural springiness of the metal tube's walls makes it curve back into its normal shape.

If you hitch up some kind of indicator needle or pointer to the free end of the Bourdon tube, you can observe its reactions to changes in fluid system pressure. By placing a properly marked or "calibrated" dial behind the indicator needle, you have a pressure gage that reads in whatever pressure units are indicated on it.

An instrument designed to measure large changes in high-pressure fluid systems will need a stiff, heavy-walled Bourdon tube. For measuring slight changes in relatively low-pressure systems, the instrument must have a more flexible and sensitive Bourdon tube.

QUIZ

1. The action of one body on another tending to change the state of motion of the body acted upon is called

2. In physics, pressure is per unit area.

3. Two systems of measuring pressure are in units like pounds per square inch or in units like .

4. Atmospheric pressure at sea level is pounds per square inch.

5. A "standard atmosphere" is of mercury.

6- A measures atmospheric pressure.

7. The word "Aneroid" means

a. Measurer.

b. Not wet.

c. Altitude.

d. Box.

8. The casing of an aneroid gage is usually .

9. The metal capsule of a rate-of-climb indicator has a
in it.

a. Lever.

b. Small hole.

c. Gas sealed.

d. Heater.

10. A liquid is almost while a gas can be easily

11. The tube is used in many fluid pressure indicating
instruments.

12. In physics, force is ordinarily measured in

13. The is the lower part of the atmosphere*

a. Troposphere.

b. Stratosphere.

c. Ionosphere.

d. Exosphere.

14. An aneroid instrument is an instrument operated by a pressure sensitive

15. The indicator is an aneroid with a leak in the capsule.

16. Instruments are operated by pressure , , or
by

17. The standard temperature at sea level is degrees
Fahrenheit.

18. Boyle's Law states that if the temperature remains constant, the pressure exerted by a
confined gas varies

19. The Bourdon tube in an aircraft instrument is made of .

CHAPTER

REMOTE READING INSTRUMENTS

THE D-C SELSYN

A pilot does not have enough time and, in most cases, is not positioned to check by personal observation whether the landing gear on his airplane has retracted properly after a takeoff. Neither can he tell by looking whether the wheels have locked securely in landing position as he prepares to approach for a landing. He does, however, have an instrument in the cockpit to inform him about the position of his landing gear. Such an instrument is called a position indicator.

Position indicators are used not only in connection with landing gear, but with wing flaps, cowl flaps, and oil-cooler shutters, as well as with some types of fuel-quantity gages. Most position indicators used in Naval aircraft are of the type known as d-c selsyn instruments. They consist of an indicator and a transmitter, and are designed to operate on direct current (d-c) from standard battery-generator electrical systems.

The transmitter unit of a d-c selsyn instrument is housed in a metal case. The unit consists of a resistance strip of near-circular or arc shape, and a movable contact arm which is linked by a lever system to the landing gear, flaps, shutters, float arm, or any like part of the aircraft whose position is to be indicated. Thus, any change in the position of the movable aircraft part linked to the transmitter will vary the resistance of the circuit.

The basic mechanism of the indicator is the indicating element. It contains a coil connected into the circuit with the variable resistance of the transmitter. Thus, when the

landing gear (for example) changes position, its motion varies the resistance in the electrical circuit. This variation affects the flow of current in the circuit, and the current change is shown on the indicator. A diagram of the three-wire d-c selsyn circuit is seen in figure 5. D-c selsyns use a magnet for the moving part. The magnet's position is controlled by the current in two or more coils (a 3-wire d-c selsyn has 3 coils, all stationary). A pointer is attached to the magnet, and the change of landing gear position will be shown on the instrument dial by this pointer.

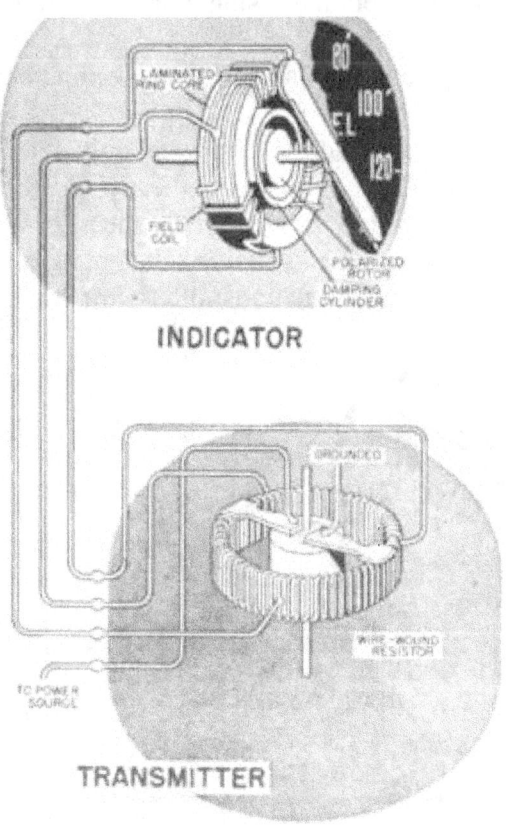

INDICATOR

TRANSMITTER

THREE-WIRE SYSTEM
The transmitter in the three-wire system consists essentially of a circular resistance winding on which a pair of diametrically opposite brushes operate. These apply direct current to the winding and are free to rotate so that current can be applied to any diameter. The brushes are carried on the end of a shaft extending outside the transmitter housing. They so rotate with the shaft that by coupling it to the motion which is desired to be remotely indicated, the brushes will follow the motion and apply voltage to the transmitter winding on a particular diameter. The toroidal resistance winding has three taps spaced 120° apart. The voltage existing between any pair of taps depends upon the position of the brushes on the winding.

The indicating element used in the three-wire system essentially consists of an annular core, a permanent-magnet rotor, a damping cylinder and three field coils. The core is made up of circular laminations of ferro-magnetic material. The three field coils are placed on the core at intervals of 120°. The leads between the coils are connected to the three taps of the transmitter winding.

Currents in two adjacent coils are in such a direction as to set up opposing fields in the ring. As a result, instead of flux going around the ring, it is forced out of the ring and must cross the inside space.

As voltage at the transmitter taps is varied, the distribution of current in the indicator coils is affected and direction of the magnetic field across the indicating element is changed. As the magnetic field changes direction, the polarized rotor follows it. Since there are no springs or restraining forces on the rotor, its movement is independent of the battery voltage. Likewise, since the coils are arranged symmetrically, the rotor position is not affected by changes in coil resistance occurring with changes in temperature.

TWO-WIRE SYSTEM

The transmitter used in the two-wire system has a single brush operating on a circular resistance winding. The indi-

eating element has two field coils, placed 120° apart and approximately 120° from a 0.015-inch slot in a laminated circular core.

This slot is to prevent flux from going around the ring when only one coil is excited, as is the case when the brush is at one extreme end of the travel. In other words, the slot offers reluctance to the flux and makes part of it go across the ring where the rotor is located.

Connections in the two-wire system are made as shown schematically in figure 6. The battery supply is connected across two taps made in the transmitter winding, and also across the two ends of the indicator field coils. The trans-

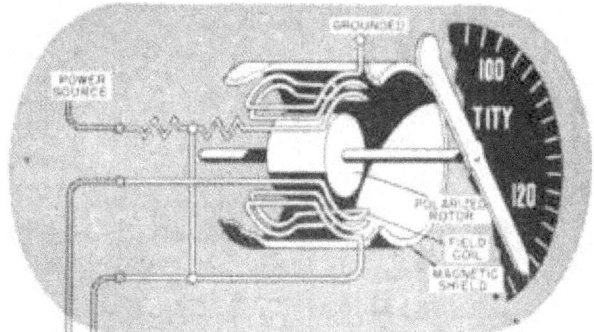

INDICATOR

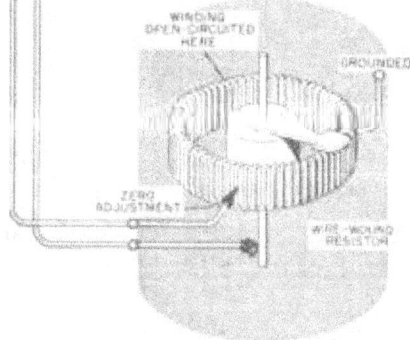

TRANSMITTER

Figure 6.—Schematic diagram of a Selsyn two-wire (d-c) system.

mitter brush is connected to the indicator at a point between the two field coils.

As the transmitter brush is rotated toward one end of the winding, the current in one of the field coils is reduced until at the very end of the winding the current becomes zero. With the transmitter brush rotating in the opposite direction, the current in the other indicator field coil is reduced. When the brush has reached the opposite end of the coil, the current will then be zero in the second indicator coil. As the brush moves from one end of the used portion of the coil to the other, the magnetic field in the indicating element shifts from a point 60° on one side of the slot to a point 60° on the other side, resulting in a total angular movement of 120°.

The portion of the winding included between the taps of the two-wire transmitters varies for different models. The transmitter winding is open-circuited by cutting several turns of the coil at some point outside its effective range.

The two-wire system requires fewer connections than the three-wire system and it is especially suitable for installations involving a pointer rotation of 90° or less.

Many sorts of "teaming'* combinations are possible with position indicators and transmitters. The indicator instru-

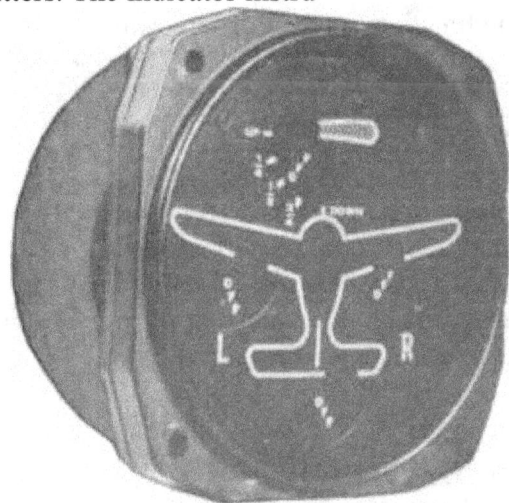

Figure 7.—Typical position indicator dial.

ment may contain several indicating elements rather than just one. Each element in such an instrument is operated by a separate transmitter. Thus you can save precious space on the instrument panel by making one dial do several jobs. A typical position-indicator team is an instrument showing landing wheel and wing flap positions on the same dial. An example of such an instrument dial is shown in figure 7.

When you're installing any position indicator unit in an airplane, ordinary conduit connectors can be used for connecting the indicator to the transmitter. The indicators are not affected by the length of the leads. Two things must always be kept in mind—official wiring diagrams must be followed closely, and the contact posts on transmitters must not be reversed. Keeping the polarity correct is vital, because reversal of the transmitter terminals will make the indicator work backwards. If the landing wheels are down, for instance, a carelessly wired indicator system might show that they are up, and vice versa. Be sure to consult the directions for each type of instrument before making adjustments. You doubtless will have to check on the linkage which connects the landing gear or like part to the transmitter, too, to see that it's exactly the right length. Otherwise the indicator will give inaccurate readings.

Maintenance of d-c selsyn indicators involves only minor repairs. If an instrument indicates incorrectly, check the external connections carefully to see that they're clean and tight, and test the leads to be certain they aren't broken.

If these checkups don't remedy the trouble, you may discover an accumulation of dirt under a rotating contact on the resistance strip in the transmitter. Such dirt should be removed from the rotating contact by means of a fine abrasive paper, and from the resistance strip by using a small brush. Don't try to repair a faulty indicator. Remove it and replace with a serviceable instrument.

AUTOSYN INSTRUMENTS

Single-engine airplanes manage very well with direct-reading gages and indicators on their instrument panels.

Some twin-engine airplanes do too. Large, multiengine airplanes, however, offer a somewhat more complicated problem with respect to instruments. Imagine the long spans of rigid or semirigid tubing, delicate capillaries, and special bracing you would need to connect direct-reading instruments from a central instrument panel to the widely separated power plants of a big four-engine flying boat, for example. The system would be complicated, and downright impractical.

A method of measuring engine and airplane functions at remote points and transmitting the measurements to a central point by electricity has been worked out. It's called the autostn system, and is used on many Naval aircraft.

What functions do autosyn instruments measure and indicate? Almost anything mechanical, or that can be made to cause mechanical action. Autosyn instruments are available to indicate the amount of fuel in tanks, the fuel pressure, oil pressure, manifold pressure, engine r.p.m., oil temperature, fuel flow rate, carburetor mixture, air temperature, and the position of such moveable parts as flaps, shutters, and landing wheels. Such functions are indicated electrically at a point remote from the place of measurement of the quantity or position, thus saving weight and greatly reducing fire hazard, mechanical difficulty and the possibility of losing fuel and oil.

The autosyn system for a particular job is made up of two units—a transmitter and an indicator. The transmitter is mounted close to the point where the function to be measured takes place, and is connected electrically to the indicator on

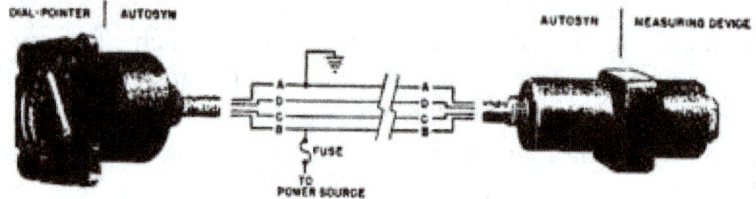

INDICATOR   TRANSMITTER

Figure 8.—Schematic diagram of a tingle function autosyn syttom.

the cockpit instrument panel. A schematic of an autosyn system for a single function is shown in figure 8.

An autosyn transmitter unit for any specific job contains a measuring device much like the working mechanism of a direct-reading instrument used for a similar purpose in a single-engine airplane. In addition it contains linkage for converting the action of the measuring device into rotary motion, and the rotary motion is transmitted to the autosyn motor.

Different types of measuring devices are used, of course, in the various autosyn transmitters. The nature of the device depends upon what the particular equipment is designed to measure. But the autosyn motor is basically the same for all transmitters, regardless of purpose.

The indicator also contains an autosyn motor, plus a pointer and dial. Because of the way in which the autosyn system works, any indicator may be converted from one purpose to an entirely different use simply by replacing the original dial with an appropriate one. Some indicators are single, others are dual. The latter type is used when space on the instrument panel

is limited, or when you want

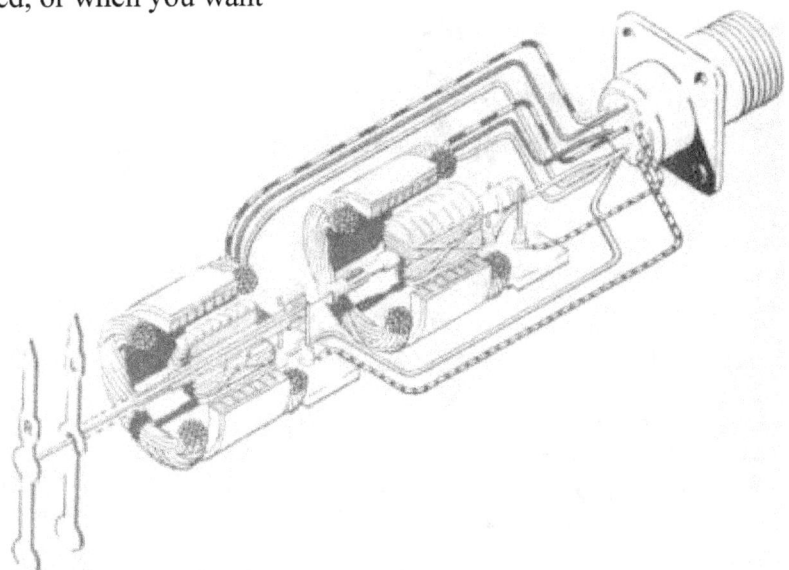

Figure 9.—Schematic wiring of a dual autosyn indicator.

to be able to compare similar functions of two engines on the same dial.

The autosyn motors of the transmitter and indicator are similar in construction and identical in electrical characteristics. They look very much like ordinary small motors, but they operate quite differently. If you're inclined to judge electric motors by their power and r.p.m., you wouldn't rate autosyn motors very high.

The most important parts of an autosyn motor are the rotor and stator. The rotor is the turning part, and its shaft is mounted on precision ball bearings. The stator is the part that stands still. You can get a better picture of what these parts look like by referring to figure 9.

The basic principle of autosyn operation is the duplication of the motion of one motor in another. The two motors can be a great distance apart and still operate in perfect synchronism. The rotor of one motor follows the slightest motion of the rotor of the other motor. Simple electric wiring between the transmitter and indicator eliminates all mechanical connections and tubing between them.

Each stator consists of a closed three-phase, two pole winding, connected in a Y. The rotor also has two poles, but has a simple, single-phase winding. Alternating current is supplied to the rotor windings, which are connected in parallel circuits. When the voltage is applied, both rotors are energized.

The stators are not connected to any power supply, but to each other (in parallel) by means of three wires, as you see in the circuit diagram, figure 9.

The magnetic field set up by the transmitter rotor creates a voltage in the stator winding—in fact, a different voltage in each of the three phases of that winding. It happens that for any given position of the rotor these three voltage values are fixed, and correspond only to that position. As the transmitter rotor is turned, the voltage in one phase of the stator winding grows larger and that in the others becomes less. Thus a new set of voltage values is set up for every possible position of the rotor.

274033 9 —54 3

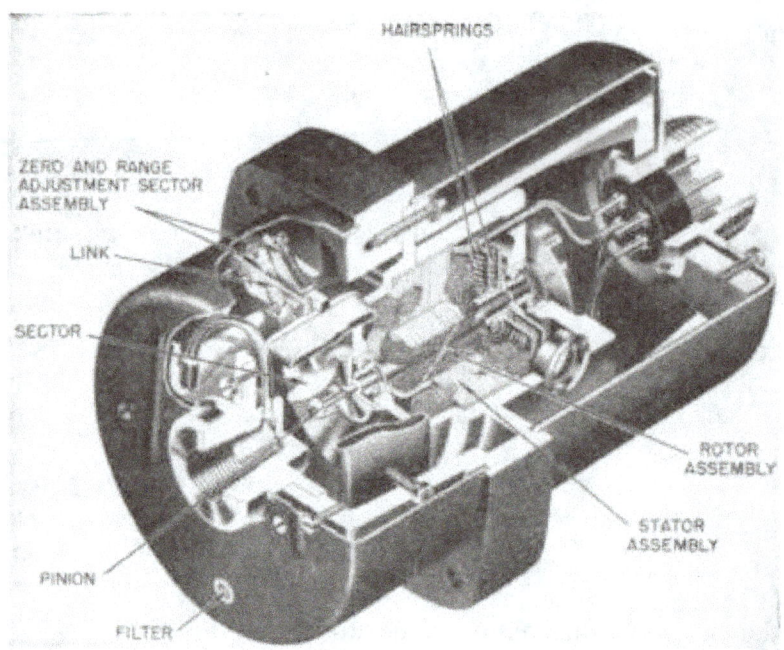

Figure 10.—Cutaway of an autosyn transmitter (oil pressure).

Since the stators of transmitter and indicator are connected electrically to each other, you have the same voltages in both stator windings. The rotor of the indicator, therefore, takes a position inside its stator that matches the position of the transmitter rotor. By placing a calibrated dial in front of the indicator rotor and attaching a pointer to the rotor shaft, you have an autosyn-type remote indicating instrument.

What sets the position of the transmitter rotor in the first place? The measuring device in the transmitter case, which you read about a few paragraphs back. The moving parts of the measuring device are attached to linkage mechanisms which, you remember, convert the movement into rotary motion. The rotary motion is what turns the transmitter rotor to position.

The alternating current is usually supplied by rotary inverters driven from the aircraft's direct current system. In some later model aircraft the main alternators are the source of supply.

It is possible to use one indicator with several transmitters, so that indications from several sources may be obtained from a single dial. This is particularly important when instrument-panel space is crowded. Also, two indicators can be driven from a single transmitter. This is desirable in airplanes which carry flight engineers in addition to pilots, and provide a separate instrument panel for the engineer's use.

TRANSMITTER
OPERATING REAR AUTOSYN (FRONT POINTER)
TRANSMITTER OPERATING FRONT AUTOSYN (REAR POINTER)

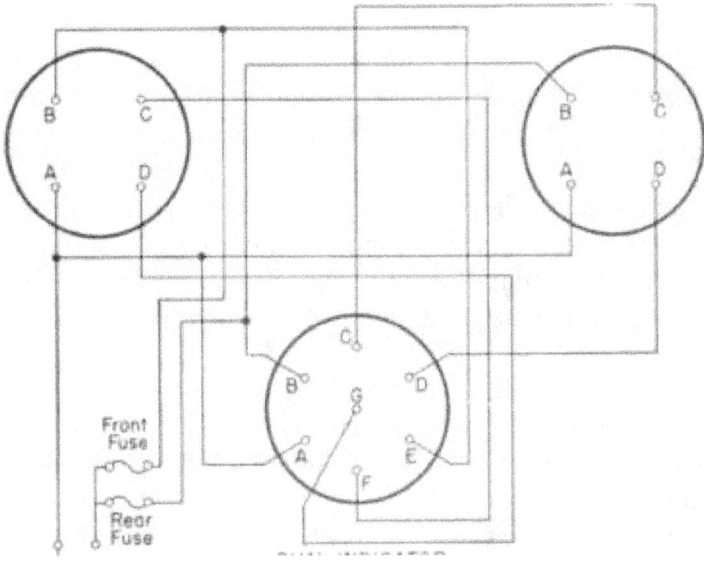

26 Volts
Ground if desired
DUAL INDICATOR
Figure 11

connected to one dual autosyn indicator.

The dual autosyn indicator is used with two transmitters, each operating one of the indicator pointers. Both pointers rotate about the same center and are read on the same dial, as shown in figure 12. The dual indicator mechanism has

Figure 12.—Dual autosyn indicator dial.

two individual motors, one mounted directly behind the other. The rotor shaft of the front motor is hollow. An extension shaft from the rear motor reaches to its dial pointer by passing through this hollow shaft. The electrical connector receptacle on the rear of a dual indicator is constructed so as to provide connections for both motors.

The autosyn power supply should in most cases be switched on before starting the airplane engines. Otherwise there may be a sudden severe displacement of the rotor positions, and you may have to recalibrate the instruments. If you are making an inspection of the autosyn instruments, idle the engine at 600 r.p.m. If indicator readings are not within the specified tolerances (error allowances) for each individual instrument, check all wiring and external connections. Use an ohmmeter to test for broken wires. Make certain that connections to and from the various transmitters and indicators are not reversed, and that the power supply input and output are steady and otherwise normal. If instruments still misbehave, they should be removed and sent to an instrument repair shop for overhauling.

If you're not definitely sure whether instrument trouble is caused by a faulty transmitter or by the indicator, test the indicator by connecting it with a spare indicator which has its cover glass removed. The spare indicator thus takes the place of the transmitter during the test. Then turn the pointer of the spare indicator by hand. If the indicator in the cockpit doesn't "follow" correctly, you know that the trouble is either in the connecting wires or in the cockpit indicator. By checking the wiring with an ohmmeter, you soon can isolate the cause of the trouble. Your station probably has autosyn test kits available for conducting these checks.

The mounting of autosyn transmitters is worth some special mention. There is, of course, plenty of severe vibration in the engine nacelles of airplanes, so it's highly desirable to place transmitters on vibration-absorbing mountings. It is usually impracticable to mount individual transmitters on separate shock absorbers because each instrument is rather small. To get around this problem, most of the various transmitters for each engine are usually grouped, and mounted together on a shock-mount panel. Exceptions to the rule are position indicator and fuel-flow transmitters, which are generally mounted individually on rubber or similar material because their external connections are rigid.

Pressure and mechanical connections to the other transmitters must be made by means of hose or flexible metal tubing so that free movement of the mounting panel will not be hindered, and so the connectors won't be loosened by vibration. Flexible lines and electrical conduits must be attached in such a way that they won't force the panel in one direction or another and thus ruin the effectiveness of the shock absorbers.

MAGNESYN SYSTEM

The magnesyn system is an electrical self-synchronous device used to transmit the direction of a magnetic field from one coil to another. In a magnesyn, a soft iron ring is placed around a permanent magnet so that most of the magnet's

lines of force pass within the ring. This ring, which may be described as a circular core of magnetic material, is provided with a single continuous electrical winding of fine wire. An electrical wiring schematic of a magnesyn system is shown in figure 13. The appearance of the winding is similar to that of the winding of the paper wrapping usually found around a new tire. Such a winding is called a toroid winding. The circular core of magnetic material and the single continuous toroidal winding are the essential components of the magnesyn stator.

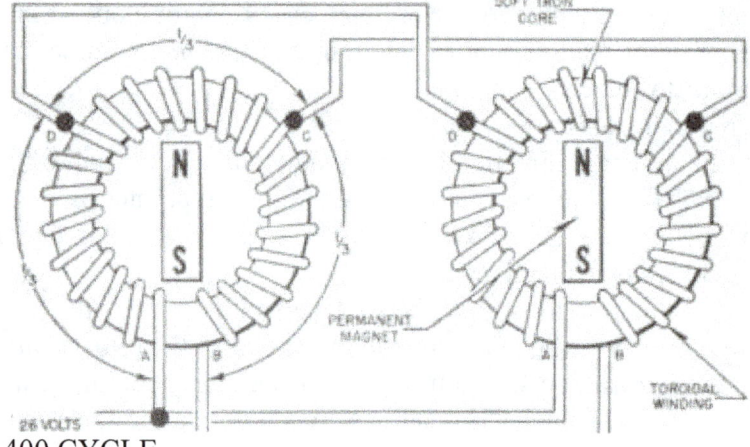

400 CYCLE

TRANSMITTING MAGNESYN INDICATING MAGNESYN

Figure 13.—Electrical wiring schematic—Magnesyn system.

The magnesyn system is essentially a method for measuring various engine functions, such as engine oil or fuel pressure, and transmitting the measurements electrically from the point of measurement to the magnesyn indicator on the instrument panel.

Magnesyn systems are used more and more as aircraft have increased in size and complexity. Methods of remote indication have become essential. An electrical system of remote indication offers a saving in weight, results in sim-

plicity in installation and maintenance, better interchange-ability of components, and reduction in military combat hazard.

QUIZ

1. Most position indicators used in the Navy are of the

type.

2. The basic element of the position indicator is the

3. The transmitter of the three-wire system consists essentially of a winding.

4. The transmitter used in the two-wire system has

a. Four brushes.

b. Three brushes.

c. Two brushes.

d. One brush.

5. The two-wire system requires connections than the three-wire system.

6. When installing position indicators, it is particularly important to keep the correct.

7. The autosyn system is composed of two units: a and an .

8. The most important parts of the autosyn motor are the and the

9. The basic principle of autosyn operation is the of one motor in another.

10. The magnesyn system is essentially a method of measuring various

11. Navy position indicator instruments consist of two units: the and the

12. The transmitter used in the system has a single brush.

13. In installing indicators, keeping the correct is vital or the indicator will work backwards.

14. The is the stationary part of the autosyn motor.

15. The magnetic field set up by the transmitter rotor creates a voltage in the

16. Current for the autosyn rotors is usually provided by a(n)

17. The autosyn power supply should in most cases be switched on the aircraft engines.

a. None of the below.

b. While starting.

c. After starting.

d. Before starting.

18. The is an electrical self-synchronous device used to transmit the direction of a magnetic field from one coil to another.

19. The system is especially suitable for installations involving a pointer rotation of 90 degrees or less,

20. If an indicator is faulty, .

a. Repair it as soon as possible.

b. Remove and replace with another instrument

c. Repair it before the next flight.

d. Throw it away.

21. The is the turning part of the autosyn motor.

22. Autosyn rotors run on current

a. A.C.

b. D.C.

c. Pulsating D.C.

d. Battery.

23. Dual indicator-mechanisms have motors.

24. A suspected indicator may be checked by connecting it to a spare with its cover glass

removed.

25. The circular core of magnetic material and the single continuous winding are the essential components of a magnesyn stator.

CHAPTER 4

PRESSURE GAGES

A military aircraft requires a considerable number of pressure gages to keep its pilot and crew informed of what is happening in various parts of the airplane and engines. One type of pressure gage indicates the pressure at which engine oil is being forced through the bearings, oil passages, and moving parts of the engine. Another type indicates the pressure at which fuel is being delivered to the carburetor. Pressure gages also measure the pressure of air in de-icer systems and gyroscope drives, of fuel mixture in the intake manifold, and of liquids or gases in a number of other systems.

ENGINE GAGE UNIT

The engine gage unit is actually three separate instruments housed in a single case. Two of these units are pressure gages. An engine gage unit is shown in figure 14. The gage unit consists of an electrical thermometer for the oil temperature, and two Bourdon type pressure gages which indicate oil pressure and fuel vented pressure at the carburetor.

The oil temperature gage is a ratiometric type galvanometer connected in a modified Wheatstone bridge circuit with an external thermocouple as one of the bridge resistances. The other resistances in the circuit are housed within the gage unit.

The tuel pressure gage indicates the pressure at which fuel is pumped into the carburetor, relative to the carburetor air pressure. The fuel pressure gage has two Bourdon tubes which are connected through a linkage to the fuel pressure pointer. One of the Bourdon tubes is actuated by the fuel pressure, the other by the carburetor static pressure. Deflec-

Figure 14.—Engine gage unit.

tions of the tubes, due to the pressures applied to them, act against each other through the linkage so that the gage gives the relative fuel pressure. Thus, if the fuel pressure at the carburetor is 28 p.s.i. above atmospheric pressure, and the carburetor static pressure is 3 p.s.i. above atmospheric pressure, then the fuel pressure gage will indicate a pressure of 25 p.s.i. The gage scale runs from zero to 35 p.s.i., calibrated in increments of 1 p.s.i.

The oil pressure gage has a single Bourdon tube arrangement operating through a linkage to actuate the oil pressure pointer. The scale range is from zero to 200 p.s.i., calibrated in increments of 10 p.s.i.

OIL PRESSURE GAGES

Oil pressure instruments are required on all aircraft engines to show the pressure at whicli the lubricant is being forced to the various points in the lubricating system. The measurement indicates to flight personnel whether or not the oil is circulating under proper pressure prior to takeoff. During flight, it warns of impending engine failure caused by exhausted oil supply, failure of the oil pump, burned-out

bearings, broken oil leads, and other causes indicated by loss of pressure. The oil pressure limits vary depending upon the particular engine.

The standard oil pressure gage has a Bourdon tube mechanism, essentially a metal tube of elliptical cross section, enclosed in a bakelite (plastic) case, as shown in figure 15. Its range is from 0 to 200 pounds per square inch (p.s.i.) and its scale is marked in graduations of 10 p.s.i. There is a single connection on the back of the case, leading directly into the Bourdon tube.

Figure 15.—Oil pressure gage.

Some oil pressure indicators are incorporated on the left dial of the engine gage unit used in some aircraft. Reciprocating aircraft engines are equipped with engine-driven oil pumps. Whenever the engine is running, oil is forced through the engine under pressure. This pressure is controlled by a pressure relief valve which can be set for the recommended pressure of the specific engine. The pressure gage is connected into the system at a point between the relief valve and the engine.

At the point where an oil pressure gage is connected into the system, there is a restriction in the line. This restriction

prevents the surging action of the oil pump from damaging the gage, keeps the gage pointer from oscillating violently through wide ranges, and makes it possible to read the gage accurately. Because of this, and the sluggishness of cold oil, the gage may fail to indicate any pressure immediately after the engine is started in cold weather; however, if the system is working properly, this will shortly correct itself.

When the gage does not respond, or indicates incorrectly, or oscillates excessively, check all tubing and all tube connections throughout the system for pressure leaks. Tighten connections and replace the tubing if necessary. If the system is pressure tight, the fault is within the instrument. It should be removed and replaced with a serviceable one. Also, if there is excessive error at zero, the instrument should be replaced.

COLD WEATHER MAINTENANCE

In order to insure accurate and dependable oil pressure gage readings during cold weather running of the engine, or when the oil gage line becomes congealed and sluggish, the following procedure should be followed.

1. Remove cap from tee in the oil pressure line in the engine accessory near the quick disconnect.

2. Remove cap from tee at the gage end of the oil pressure line. The oil gage line is identified by a one-half inch yellow stripe on the tubing.

3. Connect a pump or other means of forcing SAE 10W oil into the line at the gage end of the line. .

4. Pump oil in until all heavy engine oil is forced out and clear, and bubble-free oil runs through the instrument.

5. Cap the tee in the accessory system, remove the pump from the instrument end of the line, and replace the cap on the tee. Tighten the cap sufficiently to make certain that the flare on the cap seats properly.

For extreme low temperature operation, kerosene or an oil with a viscosity rating lower than SAE 10W should be used.

It may be necessary to repeat this procedure frequently when pressure readings become sluggish under winter weather conditions.

MAGNESYN OIL PRESSURE SYSTEM

The Magnesyn oil pressure system used on some aircraft is essentially a method of measuring engine oil pressure directly, and transmitting the measurements electrically from the point of measurement to the Magnesyn indicator on the instrument panel. The use of electrical transmission of measurement in the Magnesyn system eliminates the need for direct pressure lines from the engines to the instrument panel. Possibilities of fire'hazard, loss of oil or fuel, and mechanical difficulties are greatly minimized. The system requires a 26-volt, 400-cycle source of power. The Magnesyn system will operate satisfactorily within a temperature range of +71° C. (+ 159.8° F.) to -54 C. (-65° F.) A temperature as low as —65° C. (— 85° F.) will not have a permanent adverse effect on the instruments.

The Magnesyn system is composed of two main units—the Magnesyn transmitter (bottom) and the Magnesyn indicator (top) as shown in figure 16. A transmitter is provided for each engine but is usually mounted on a bracket attached to the engine between the oil cooler and the air duct adapter near the top of the engine. An oil line connected to the engine oil system supplies pressure to the transmitter. Electrical current corresponding to the pressure of the system is relayed to the oil pressure gage (indicator) located on the instrument panel. The electrical leads between the transmitter and the indicator may be any reasonable length without noticeable effect on the indication.

Oil pressure Magnesyn transmitters are composed of two main parts—a bellows type mechanism for measuring pressure and a Magnesyn assembly. The pressure of the oil causes linear displacement of the Magnesyn magnet. The amount of displacement is proportional to the pressure. Varying voltages are set up in the Magnesyn stator, depending on the position of the magnet. These voltages are transmitted to the magnesyn indicator which indicates on a dial the value received from the transmitter.

The Magnesyn indicator consists essentially of a Magnesyn, a pointer, and a graduated dial. The pointer is at-

FUEL TRANSMITTER OIL TRANSMITTER

Figure 16.—Schematic of a dual Magnesyn system.

tached to the shaft of the magnet. Rotation of the magnet causes the pointer to move, indicating the value received from the transmitter. The indicator contains no brushes or sliding contacts. The only source of friction lies in the pivot jewels in which the rotor shaft (magnet) turns. The use of jewel pivots and a light rotor reduces friction to a minimum.

In some cases dual indicators are used to obtain indications from several sources when space is greatly reduced on the instrument panel. Each dual indicator is connected to

two transmitters and indicates, by means of two pointers on a single dial, the values received from the two transmitters. Thus, four Magnesyn transmitters will indicate measurements on two Magnesyn dual indicators.

MAGNESYN ASSEMBLIES

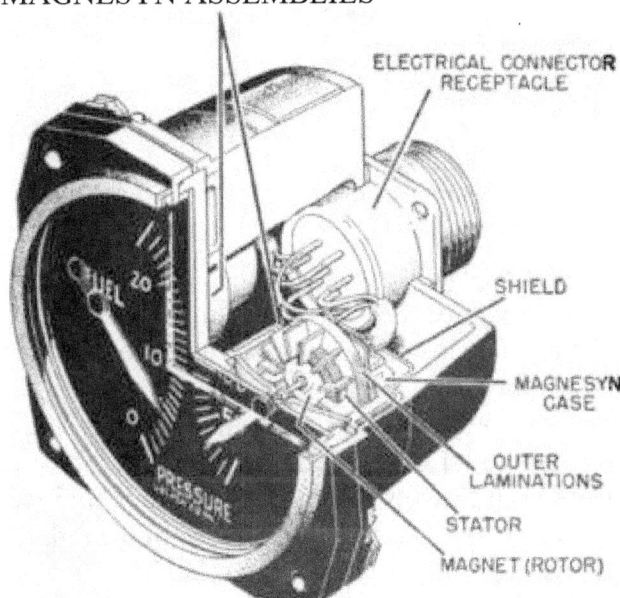

Figure 17.—Cutaway of Magnesyn oil and fuel pressure indicator.

On some aircraft, both oil and fuel pressure transmitters are joined through a junction box and operate a Magnesyn oil and fuel pressure indicator (dual side-by-side) thus combining both gages within one case as shown in figure 17.

INSPECTION AND MAINTENANCE

Prior to each flight the Magnesyn oil pressure system should be checked for proper operation. With the engine or engines running, the indicator readings should be consistent with engine conditions.

The electrical wiring should be checked on periodic inspections for proper insulation, continuity, and anchorage. The electrical connector plugs should be checked to insure there are no broken or loose connections and leads. After

1000 hours of operation, the oil pressure transmitter and indicator should be removed from service and sent to instrument overhaul. Ordinarily, the operation of the system is automatic after it is installed. If malfunctioning occurs, first check to determine if the instrument or the installation is at fault. If the instrument is faulty, remove and replace it with a serviceable unit.

Trouble in the oil pressure Magnesyn system usually falls within three groups; trouble in the power supply; trouble in the Magnesyn units; and trouble in the electrical or mechanical connections to the Magnesyn units. To localize the source of trouble, each group must necessarily be checked.

FUEL PRESSURE GAGES

The fuel pressure system measures and indicates the difference between the air and fuel pressures at their respective inlets to the carburetor. It indicates the uninterrupted flow of fuel to the carburetor(s) while switching from one fuel tank to another. Before takeoff, it indicates whether or not fuel is being supplied to the carburetor(s) steadily under the proper pressure. This is a check on the functioning of the fuel pump and, to a certain extent, on the entire fuel system.

In aircraft using reciprocating engines, the fuel pressure gage on the instrument panel is a member of the Bourdon tube family. Two tube nipples are provided in the back of

Figure 18.—Front and rear of fuel pressure gage.

the case—one marked "P," and the other marked "V," as you see in figure 18, to connect the gage to the engine fuel system and to the air. The instrument is Chus adaptable for use with both internally and externally supercharged engines.

On some internally supercharged engines, only the "P n connection is made. The air vent remains open. On externally supercharged engines and some internally supercharged types, you connect the "V" vent to the air pressure chamber of the supercharger. On engines equipped with a carburetor ram, the vent U V" may be connected to the carburetor because of the ram pressure. In any case you're getting the same kind of reading—the difference between the air pressure and fuel pressure as they enter the carburetor.

You find that with pressure-discharge carburetors the range of the fuel pressure gages used is from 0 to 25 p.s.i. With all other types of carburetors, however, the gages used have a range of from 0 to 10 p.s.i. On either instrument, the dial is marked in steps of 1 p.s.i.

With engines using pressure-discharge carburetors, the normal pressure gage reading during engine operation should be from 15 to 17 p.s.i. With other types, 6 to 8 p.s.i. is normal. The fuel pressure gage is a sturdy instrument and will stand a certain amount of overpressure— usually about one-half of its calibrated range—without serious effects. If you use the fuel system hand pump too violently, it is possible to build up excessive pressures beyond the limits of the relief valve, and damage to the fuel pressure gage will result. Warning— watch the gage while hand-pumping fuel, and stay out of gage trouble!

DE-ICING PRESSURE GAGES

The rubber expansion cells, which de-ice the leading edges of wings and stabilizers, are operated by a compressed air system. The de-icing system pressure gage measures the difference between prevailing atmospheric pressure and the pressure inside the system, indicating whether there is sum-

cient pressure to operate the expansion cells. The gage also provides a method of measurement when you're setting the relief valve and regulator of the de-icing system.

This instrument's case is made of bakelite (plastic) and is pierced with a vent at the bottom to keep the inside at atmospheric pressure as well as to provide a drain for any moisture which might accumulate. The pressure-measuring mechanism consists of a Bourdon tube and a sector gear with a pinion for amplifying the motion of the tube and transferring it to the pointer. The de-icing system pressure enters the Bourdon tube through a connection at the back of the case.

The range of de-icing pressure gages is from 0 to 20 p.s.i., with the scale marked in graduations of 2 p.s.i., as shown in figure 19. The pointer, numerals and main graduation

Figure 19.—Dial of a de-icing pressure gage.

marks are coated with luminous paint. When installed and connected into an airplane de-icing pressure system, the gage reading always remains at zero unless the de-icing system is operating. When the system is in use, the gage pointer will fluctuate from 0 p.s.i. to approximately 8 p.s.i. under normal conditions, because the expansion cells are periodically in-flated and deflated. Don't confuse this normal fluctuation with oscillation, which is not a normal condition in any gage and must be corrected when it occurs.

SUCTION GAGES

Suction gages are used on aircraft to indicate the amount of suction that actuates air-driven gyroscopic instruments. It is particularly important in instrument flying that all the flight instruments on the instrument board function properly. Since any reduction in the normal amount of suction which actuates the gyroscopic flight instruments will impair proper indication, the suction gage is the means provided to inform the pilot continually of the exact amount of this suction.

The spinning rotors of gyroscope instruments are kept in motion by streams of air which are directed against the rotor vanes. These air streams are not produced by pumping air into the gyroscope instrument cases. Instead, air is pumped out of them by the vacuum pump system. Atmospheric pressure then forces air into the cases through filters, and it is this incoming air that is directed against the rotor vanes to turn them.

You need a suction gage to tell whether the vacuum pump system is working properly. A suction gage case is vented to the atmosphere or to the line to the air filter, and contains a pressure-sensitive capsule or diaphragm, plus the usual multiplying mechanism which amplifies the movement of the capsule's side and transfers it to the pointer.

The pressure-sensitive capsule in a suction gage is very much like an aneroid capsule. It is not permanently evacuated and sealed, however, but is connected to the vacuum pump system through the back of the case—in much the same way that the Bourdon tube in a pressure gage is connected into a pressure system. The reading of a suction gage indicates the difference between atmospheric or filter system pressure and the reduced pressure in the vacuum pump system—in inches of mercury.

The range of the gage is from 0 to 10 inches of mercury

(Hg.), the dial being graduated in steps of one-fifth inch Hg., as you see in figure 20. The pointer hand, numerals and "inch" graduations are painted with luminous material which makes them legible for ordinary night flying.

On standard installations, you connect the gage into the gyro horizon indicator, because this instrument's air consumption is greater than that of the other gyro instruments. Hence, if the vacuum pump system is operating the gyro horizon indicator gyro properly, the reduction in pressure will be sufficient to run the other gyros as well. You'll find that the normal suction gage reading is between 3.75 and 4.25 inches of Hg.

When you're connecting the suction gage into the vacuum system, clean the threads on the nipple and apply antiseize compound. Tighten the connecting nut until the nipple seals properly, producing a tight joint but placing no excessive strain on the instrument.

If the suction gage is to be mounted on a vibration proof instrument panel and connected into another instrument on the same panel, the connection tubing needn't be flexible. But if the connecting tube is to be fastened to a rigid part of the airplane, a flexible connection should be used.

Figure 20.—Suction gage.

HYDRAULIC PRESSURE GAGES

The gear for raising the main wheels, tail wheels, flaps, bombbay doors, and certain other units is operated by a hydraulic system in many military airplanes. A pressure gage to measure the differential pressure in the hydraulic system has the important job of keeping you informed on how this system is functioning. The aircraft hydraulic pressure gages are designed for use in the airplane's hydraulic system to indicate the pressure of the complete system or the pressure of an individual unit in the system.

The bakelite case of this gage contains a Bourdon tube and a gear-and-pinion mechanism by which the tube's motion is amplified and transferred to the pointer.

The pressure to be measured enters the Bourdon tube through the pressure part of the gage. The movement of tha tube is transmitted to the mechanism which amplifies and transfers the motion to the pointer. The position of the pointer on the calibrated dial indicates the pressure being measured in pounds per square inch. The gage scale ranges from 0 to 2000 p.s.i. and is marked in graduations of 200 p.s.i., as shown in figure 21. The pointer and dial mark-

Figure 21.—Dial of a hydraulic pressure gage.

ings are luminous and can usually be seen without extra lighting during night flights.

The pumps which supply pressure for the operation of an airplane's hydraulic units are driven either by the airplane engine or by a separate electric motor. Some installations employ a pressure tank or "accumulator" to maintain hydraulic pressure at all times. In such cases the pressure gage will register continuously. With other installations, however, operating pressure is built up only when needed, and pressure will register on the gage only during these periods.

The pressures of hydraulic systems vary for different models of airplanes. In general, pressure ranges are increasing with the later models of military aircraft. Some planes have systems with pressure ranges as high as 3000 p.s.i. Since high pressures are involved, heavy-walled tubing is necessary for the connecting lines to gages.

Oscillation or noticeable changes in dial reading when no unit is in operation indicates a malfunction in the system. A snubber is usually screwed into the pressure gage and prevents pressure surges from damaging the hydraulic pressure gage.

MANIFOLD PRESSURE GAGES

The manifold pressure GAGE is used on aircraft to indicate the absolute pressure in the intake manifold of the internal combustion engine to which it is connected.

In order that the larger engines operate at various altitudes with maximum efficiency, it is necessary to provide a definite amount of oxygen to mix with the fuel for the most efficient explosive mixture. At altitudes where air is less dense, some means of forcing more air into the engine must be employed to compensate for this loss in density. To effect this, a centrifugal air pump, termed a supercharger, is employed. It is operated by the engine, either through direct mechanical coupling or by driving a turbine with the exhaust gases as in the "turbo-supercharger." By forcing air into the intake manifold of the engine with this supercharger, sea level atmospheric conditions at the higher altitudes can be approximated within certain limits.

In the lower limits of the atmosphere, care must be exercised to control the degree of supercharging. If it is permitted to become too great, the explosive charge entering the cylinder may be of sufficient quantity to cause severe damage and even structural failure of the engine. The manifold pressure gage indicates to the pilot or flight engineer at all times the pressure present in the intake manifold.

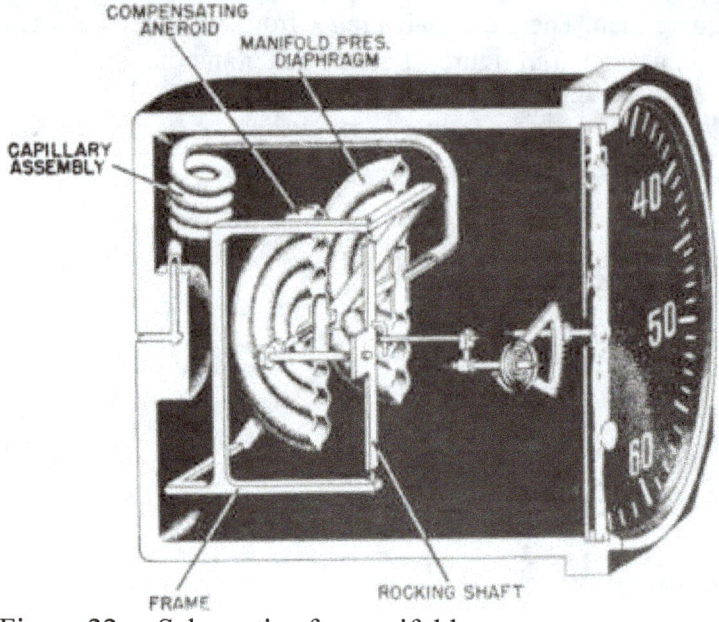

Figure 22.—Schematic of a manifold pressure gage.

Manifold pressure is measured in inches of mercury (in Hg.), because in the operation of the engine there occur pressures above and below standard atmospheric pressure, and pressures below normal are always indicated in inches of mercury. Normal atmospheric pressure at sea level (zero altitude) is approximately 30 inches of mercury.

A large hand revolving about the center of the dial indicates inches of mercury.

One type of manifold pressure gage indicates absolute pressures within a range of 10 to 70 inches Hg.; another type has a range of 25 to 65 inches Hg.; another type has a

range of 10 to 50 inches Hg.; and still another has a range of 10 to 75 inches Hg. Some have a small hand located in the upper center of the main dial which makes one complete revolution for each 10 inches of mercury pressure on the large hand. Still another type is a dual manifold pressure gage that may be used on multiengined aircraft. A schematic of a manifold pressure gage is shown in figure 22.

The readings you get on a manifold pressure gage depend on engine speed and supercharger characteristics. One kind of supercharger—the internal or geared type—is in the induction system between the carburetor and the cylinder intake ports. With this supercharging system, the gage registers the pressure between the supercharger outlet ancf the cylinder intake ports. When an external or exhaust-driven supercharger is added to the system, the carburetor is between the supercharger and the cylinder intake ports. In this case the gage is connected into the system at a point just before the fuel mixture enters the engine.

The normal position of the hands, when the airplane is on the ground with the engine(s) not running, should correspond with the barometric pressure at that point. Because of this, the sensitive altimeter may be used to check the correctness of the reading. When the engine is idling, the butterfly valve in the carburetor is not sufficiently open to permit sufficient air to flow into the manifold to fulfill the demands of the cylinders and still maintain atmospheric pressure.

Tinder these conditions, the pressure in the intake manifold is below normal. With the engine "revved-up" the butterfly-valve of the carburetor opens to admit free passage of air and the supercharger pulls in greater quantities, forcing it into the engine to increase the pressure in the intake manifold. As described above, this pressure must not exceed a specified amount for the safe operation of the engine.

The outer shell of the gage which protects and contains the mechanism has an opening at the back and provides for the connection to the manifold of the engine.

The gage contains an aneroid diaphragm and a linkage for transmitting the motion of the diaphragm to the pointer.

The linkage is completely external to the pressure chamber,, and thus is not exposed to the corrosive vapors of the manifold. The pressure "existing in the manifold enters the sealed chamber through a damping tube, which is a short length of capillary tubing at the rear of the case. This damping tube, acting as a safety valve, prevents damage to the instrument which would be occasioned by engine backfire. The sudden surge of pressure caused by backfire is. considerably reduced by the restricted capillary tubing.

The manifold pressure gage indicates the manifold pressure immediately before the cylinder intake ports. The gage is sensitive, but it is built ruggedly and will stand up under the vibration it normally gets in regular use.

What should the gage readings be under flying conditions?" That depends on the type of engine in your airplane. The correct manifold pressures for takeoff, climbing, diving, and cruising will vary with each kind of power plant.

In installing a manifold pressure gage on the instrument panel, make sure the pointer is vertical when registering 30 inches of mercury. A length of flexible tubing about 10 inches long should be connected between the instrument and the rigid tube which extends to the connection point on the manifold.

Remember .that, when an engine is not running, the manifold pressure gage reading should be the same as local barometric pressure. Check it against a good barometer. And there's a good one right in the cockpit^-your altimeter.. With your airplane on the ground, set its altimeter hand to zero, and tap the instrument panel lightly a few times to remove frictional errors. There's a barometer scale on the-altimeter, and with the altitude hands at zero this scale will show local atmospheric pressure in inches of mercury. The manifold pressure gage should agree with this pressure.

Movement of the pressure gage pointer should be slow and steady. Rapid movement or oscillation is a good sign that the damping adjustments need a check, or that there's a leak in the line or mechanism.

If the pointer fails to respond, the mechanism is in all probability defective. The gage should be removed and sent

for overhaul. If the pointer indicates incorrectly, there may be moisture in the system, obstruction in the lines, a leak in the system or mechanism, or the mechanism may be defective.

To remedy, operate the engine at idling speed and open the drain valve located near the gage for a few minutes. To clear an obstruction, blow out the lines. Any leak in the system or mechanism must be located and repaired. The mechanism can be checked for leaks by disconnecting the line at the engine end and applying air pressure until the gage indicates 50 inches Hg. Then close the line. A leak is present if the gage pointer returns to atmospheric pressure. If no leak is found, but a leak is evident, the gage should be removed for overhaul. If the pointer does not indicate atmospheric pressure when the engine is not running, there is probably moisture in the system or the gage should be recalibrated.

To check on proper adjustment of the damping restrictor, release the pressure suddenly when the gage reads 60 inches Hg. The pointer should reach 32 inches Hg. in not less than 1 y 2 seconds, and not more than 2y 2 seconds.

At first evidence of excessive pressure fluctuations or other symptoms of malfunction, check to see that the drop lines at the automatic power control unit are not reversed. The gage should be removed and tested if the lines are correctly installed.

MAGNESYN MANIFOLD PRESSURE GAGE

The MAGNESYN manifold PRESSURE system is essentially a method of measuring the absolute pressure of the fuel and air mixture as it enters the intake manifold and transmitting the measurement electrically from the point of measurement to the Magnesyn indicator on the instrument panel. The Magnesyn differential pressure system is a method of measuring the differential pressure between a high pressure (not to exceed 32 inches Hg.) and a low pressure (not to fall below approximately 10 inches Hg.). The use of electrical transmission eliminates the need for a direct pressure line

from engine to instrument panel. Possibilities of fire hazard and mechanical failures are greatly minimized. The systems require a 26-volt, 400 cycle source of power.

The Magnesyn pressure system is composed of two main units; the Magnesyn pressure transmitter and the Magnesyn pressure indicator. The transmitter is mounted close to the pressure takeoff and is electrically connected to the panel-mounted indicator. The leads between transmitter and indicator may be of any reasonable length without noticeable effect on the indication.

The Magnesyn pressure transmitter is composed of an instrument mechanism, for measuring manifold pressure, and a Magnesyn. The absolute pressure of the fuel and air mixture causes a rotary displacement of the Magnesyn magnet, the amount of displacement being directly proportional to the absolute pressure. Varying voltages are set up in the Magnesyn stator, depending on the position of the magnet. The voltages are transmitted to a Magnesyn indicator which indicates on a dial the values received from the transmitter.

The single indicator consists essentially of a Magnesyn, a graduated dial, and a pointer. The pointer is attached to the Magnesyn magnet. Rotation of the magnet causes the pointer to move, indicating values received from the transmitter.

If the indicator-transmitting manifold pressure system shows signs of malfunctioning, it will be necessary to localize the source of trouble and remove the cause. Transmitter malfunctioning is usually shown by inconsistent indicator readings. Therefore, before checking the transmitter make sure that the mating indicator and the electrical connections are working correctly.

WATER PRESSURE GAGE

The Magnesyn water pressure system is essentially a method of measuring the pressure of the water and alcohol mixture as it enters the engine and transmitting the measurement electrically from the point of measurement to the Magnesyn indicator on the instrument panel.

The Magnesyn water pressure system is composed of two main units; the Magnesyn water pressure transmitter and the Magnesyn water pressure single indicator. The transmitter is mounted close to the pressure takeoff and is electrically connected to the panel-mounted indicator.

The Magnesyn water pressure transmitter is composed of a bellows-type instrument mechanism for measuring pressure and a Magnesyn assembly. The pressure of the water and alcohol mixture causes a linear displacement of the Magnesyn magnet, the amount of displacement being directly proportional to the pressure. Varying voltages are set up in the Magnesyn stator, depending on the position of the magnet The voltages are transmitted to a Magnesyn indicator which indicates on a dial the values received from the transmitter.

The single indicator consists essentially of a Magnesyn assembly, a graduated dial, and a pointer. The pointer is attached to the Magnesyn rotor shaft. Rotation of the magnet causes the pointer to move, indicating values received from the transmitter.

Both transmitter and indicator contain an electrical unit called a Magnesyn. This unit consists of a fixed stator and a movable permanent magnet. The motion of the magnet in the indicator is rotary within the bore of the stator; in the . transmitter it is linear.

PRESSURE TORQUEMETER

The Magnesyn torque pressure system is a method of measuring the engine torque pressure and .transmitting the measurement electrically from the engine-nose point of measurement to the Magnesyn indicator on the instrument panel. Flight personnel are thus provided with a means for obtaining continuous indication of the actual power delivered to the propeller shaft. This information is valuable in reducing fuel consumption and extending the useful range of flight operation. It is, therefore, especially helpful for long-range operation of large military aircraft. The use of electrical transmission eliminates the need for a direct, oil-filled pressure line from engine to instrument panel.

MAGNESYN
ASSY.

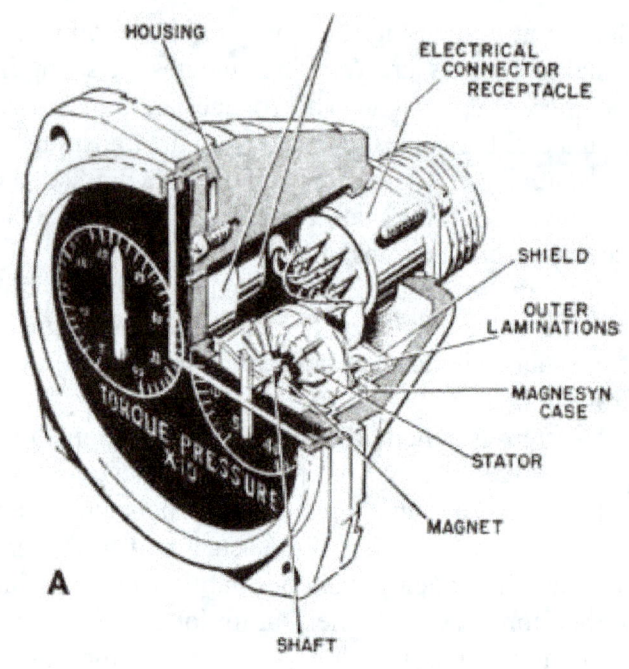

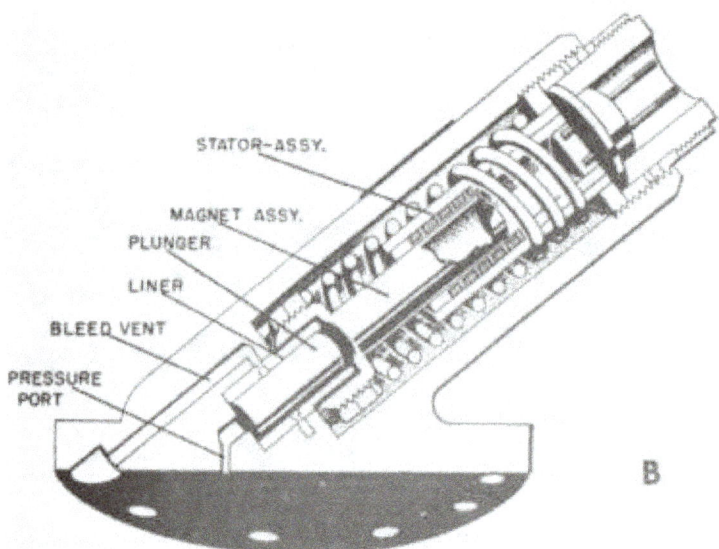

Figure 23.—Pressure torquemeter.

The Magnesyn torque pressure system is composed of two main units—the Magnesyn torque pressure indicator and a Magnesyn torque pressure transmitter. The indicator is shown in figure 23A. The transmitter, figure 23B, is mounted on the engine nose and is electrically connected to the panel-mounted indicator.

The Magnesyn torque pressure transmitter is composed of an instrument mechanism for measuring oil pressure and a Magnesyn. Pressure of the oil causes a linear displacement of the Magnesyn magnet, the amount of displacement being directly proportional to the pressure. Varying voltages are set up in the Magnesyn stator, depending on the position of the magnet. The voltages are transmitted to a Magnesyn indicator which indicates on a dial the values received from the transmitter.

The dual indicator consists essentially of two Magnesyns, two separate sets of dial markings on one dial face, and two pointers. Each pointer is attached to the magnet of one of the

Magnesyns. Rotation of the magnets causes the pointers to move, indicating the values received from the two transmitters to which the dual indicator is electrically connected.

The Magnesyn torque pressure indicator dial has been calibrated to read the oil pressure in pounds per square inch. From this reading, the brake horsepower and the brake mean effective pressure can be determined if complete data for the engine is available.

QUIZ

1. The standard oil pressure gage has a tube mechanism.
2. When an oil pressure gage indicates improperly, check first for in the system.
3. The use of of measurements in the magnesyn system
eliminates the need for direct pressure lines from the engines to the instrument panel.
4. The magnesyn indicator consists essentially of a magnesyn, a , and a
5. With pressure-discharge carburetors the range of the fuel pressure gage is from to p.s.i.
6. The rubber expansion cells for de-icing the wings are operated by a(n) system.
a. Flux gate.
b. Electric vibrator.
c. Hydraulics.
d. Compressed air.
7. Suction gages are used to indicate the amount of suction that • actuates instruments.
8. When connecting the suction gage to the vacuum system, clean the threads and apply
9. The manifold pressure gage is used to indicate —
a. Absolute pressure in the exhaust manifold.
b. Atmospheric pressure in the exhaust manifold.
c. Absolute pressure in the intake manifold.
d. Atmospheric pressure in the intake manifold.
10. Manifold pressure is measured in —.
11. The fuel pressure gage indicates the pressure at which fuel is pumped into the
12. The oil gage line is identified by on the tubing.
13. The electrical leads between the magnesyn transmitter and indicator
a. May be of any reasonable length.
b. Must be less than 15 feet long.
c. Have definite limits.
d. Must conform exactly to the drawing.
14. The use of and a light rotor in a magnesyn reduces any
friction to a minimum.
15. When should oil pressure transmitters and indicators be overhauled?
16. With all carburetors except pressure-discharge carburetors the range is from 0 to p.s.i.
17. Pressure-discharge gages normal reading should be p.s.i.
18. Normal suction gage reading is between inches of mercury.
a. 4.25-4.75.
b. 3.25-3.75.
c. 2.25^3.25.
d. 3.75-4.25.
19. Absolute intake manifold pressure is Indicated by the gage.
20. When an engine is not running, the manifold pressure gage reading should be
21. The engine gage unit is instruments in one case.
a. One.

b. Two.

c. Three.

d. Pour.

22. Before flight with engine running, the magnesyn readings should with engine conditions.

23. The range of fuel pressure gages of pressure-discharge carburetors is from to p.s.i.

24. Normal pressure gage reading for carburetors other than pressure-discharge should be from

a. 0-25 p.s.i.

b. 0-10 p.s.i.

c. 15^17 p.s.i.

d. 6-8 p.si.

25. A gage is needed to indicate whether the vacuum pump system is working properly.

26. in dial reading when no unit is in operation indicates a malfunction in the system.

CHAPTER

AIR PRESSURE MEASURING INSTRUMENTS

THE PITOT-STATIC TUBE

Of necessity, being actuated by the atmosphere outside of the aircraft, three of the most important flight instruments are connected to the pitot-static tube in order to indicate the speed, altitude, and the rate of climb of the aircraft. By

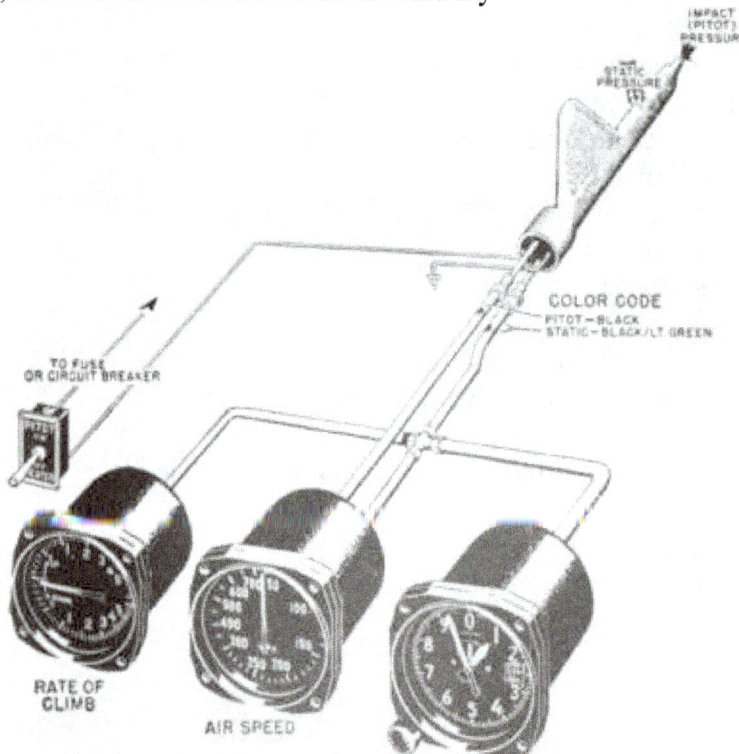

Figure 24.^—Pressure measuring instruments.

274033°—54 5 57

referring to figure 24, you can see the interrelationship of the pitot-static tube, the airspeed indicator, the altimeter and the rate-of-climb indicator.

The static tube of the pitot-static system furnishes quiet, undisturbed air at actual atmosj)heric pressure to the altimeter and the rate of climb indicator. It also supplies quiet air to the instrument case of the airspeed indicator. The static tube, however, is but one-half of the pitot-static system.

The pitot part of the system has one specific function—to receive the impact pressure of the aikstrkam as the airplane moves through the atmosphere, and to transmit this pressure accurately to the airspeed indicator.

The exterior end of the pitot-static system consists of an airspeed tube unit, mounted on the outside of the airplane at a point where the air is least likely to be turbulent, and pointing in a forward direction parallel to the airplane's line

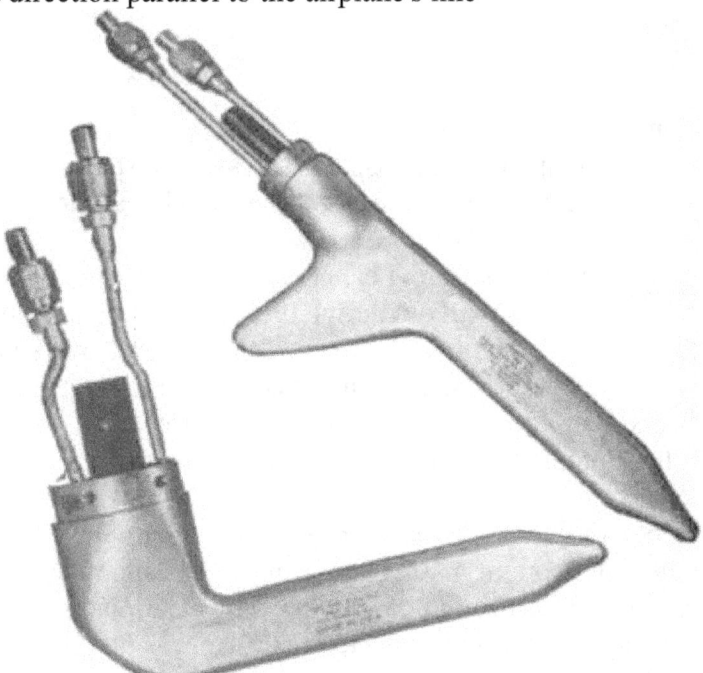

Figure 25.—Airspeed tubes for fuselage nose mounting ond for mounting on the leading edge of the wing.

of flight. One general type of airspeed tube is designed for mounting on a streamlined mast extending below the nose of the fuselage. Another type is designed for installation on a boom extending forward of the leading edge of the wing. Both types are shown in figure 25. Although there is a slight difference in their construction, actually they operate identically.

If you refer to figure 26, you'll see a diagram of the inner works of an airspeed tube. The tube is made of brass or chromeplated copper tubing, and is divided into two sections. The forward section is open at the front end so that it will get the full force of the impact air pressure. At the back of this section there's a baffle plate protecting the pitot tube from moisture and dirt that might otherwise be blown into it. There's a small drain hole at the bottom of this forward section, too, so that moisture can escape.

The pitot—or pressure—tube leads backward to a chamber in the "shark fin" projection near the rear of the assembly. A riser—or upright tube—leads the air from this chamber through tubing to the airspeed indicator, and provides additional insurance against the fouling of the system with ice, dirt, or water.

The rear—or static—section of the airspeed tube is pierced by small openings on the top and bottom surfaces. These openings are designed and located so that this part of the

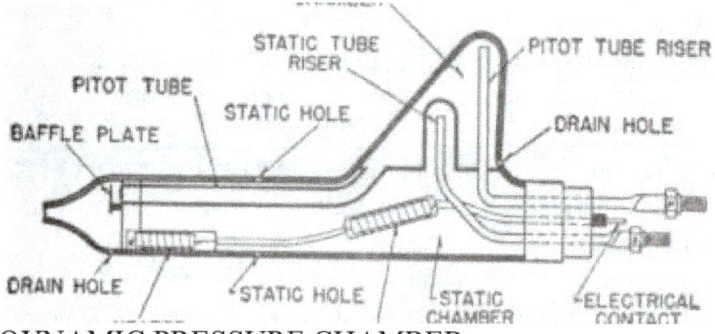

OYNAMIC PRESSURE CHAMBER .
HEATER 100 WATT
CHAMBER HEATER
ELECTRICAL CONTACT

Figure 26.—Sectional diagram of airspeed tube.

system will provide accurate measurements of atmospheric pressure in static—or still—condition. The static section's main chamber also contains a riser tube, which leads to the airspeed indicator, the rate of climb indicator, antf the altimeter.

Some airspeed tubes are provided with two heating elements to prevent icing during flight These elements are sturdy and not likely to burn out during the life of the tube. Under ice-forming conditions during flight, the heating elements can be turned on by means of a switch in the cockpit. The electrical circuit which operates these elements is connected through the ignition switch for all engines, so that if you forget to turn off the heater switch there will be no drain on the airplane's battery system when the engines aren't running. This safety device also serves to lengthen the life of the heating elements.

There is a union connection in both lines from the airspeed tube, close to the point at which the tube is attached to the mounting boom or mast. These connections simplify removal and replacement, and are reached through an inspection door in the wing or fuselage. If you're removing an airspeed tube, these unions should be disconnected first^-before you remove the mounting screws and lock washers.

You must be careful about keeping airspeed tubes clean at all times. When an airplane is not flying, its airspeed tube should be covered by a sack made of cloth or leather. Test the electrical heater circuit for satisfactory connection. The voltage drop across the heater should be about 24 volts when it's in operation.

When the static line is properly connected to all the instruments it serves, the whole static system should be tested for leaks. The altimeter will provide you with a good testing standard. Connect the static airholes of the airspeed tube to a vacuum pump, and set the altimeter pointers to zero. Apply the vacuum slowly until the altimeter indicates 1,000 feet of altitude and then pinch off the vacuum source. The altimeter pointers should not change position by more than 150 feet in 1 minute. Otherwise there's a leak. Warning.

Don't apply pressure to static lines. You'll wreck the instruments if you do!

To test the pitot—or impact pressure—line, seal off the pressure system drain holes with masking tape. Connect the pitot pressure opening of the airspeed tube to a source of air pressure—and apply the pressure slowly until the airspeed indicator point registers 150 knots. Then pinch off the pressure source. The pointer shouldn't drop more than 10 knots in 1 minute.

Kemember^o tap the instrument panel to remove friction effects during tests of either static or pitot lines. Otherwise you'll get faulty instrument readings. And never apply vacuum to the pitot pressure line when making tests. That's as taboo as shooting pressure in the static system.

AIRSPEED INDICATORS

Every airplane has a certain range of flying speeds at which it operates with greatest efficiency and safety. This range lies between the lowest speed at which it will not stall and the highest speed at which it does not dive dangerously. How does a pilot keep advised of an airplane's rate of travel through the air in which it flies? He refers to the airspeed indicator.

This instrument has a variety of useful information to give to flyers. Its readings are important when you're estimating ground speed, bomb sighting, or aiming aerial guns. It helps in determining throttle settings for the most efficient flying speed, and provides a basis for calculating the best climbing and gliding angles. It warns you when diving speed approaches safety limits as regards your airplane's structure. And—since airspeed increases when the nose drops, and decreases in a climb—the airspeed indicator is an excellent check on whether you're maintaining level flight.

Measures Pressure Differences

As you learned in reading about the pitot-static system, an airspeed indicator is fed with air at two different pressures during flight. Air at a pressure built up by the, forward speed of the airplane is led from the pitot tube into a diaphragm capsule inside the instrument case. Air at undisturbed atmospheric pressure is led from the static tube into the case itself. The difference between these two pressures, therefore, depends on the forward speed of the airplane, and is measured by the expansion and contraction of the capsule. In figure 27 you can see the working parts of a typical airspeed indicator. The connections (P) and (S) are for the pitot line and static line, respectively. The capsule (A) in center of illustration is linked to the pointer mechanism by a system of levers and gears. The case (K) is airtight except for the static line connection.

SECTOR

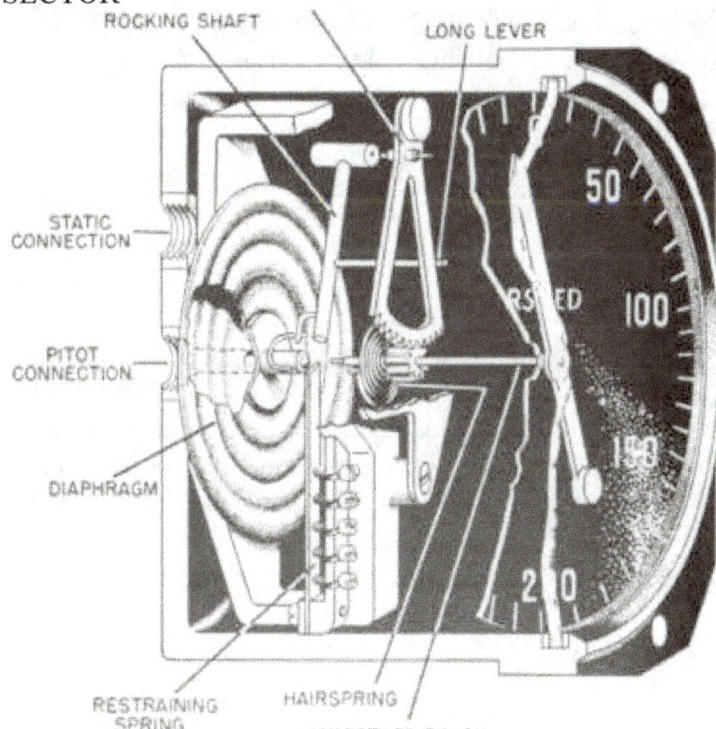

Figure 27.—Mechanism of an airspeed indicator.

One type of airspeed indicator used by the Xavy has a range from 0 to 160 knots. This type is used principally in trainer planes. It is less sensitive than—and lacks the accuracy of—the 430-knot type installed in the Navy's airplanes of the line. The dial of a 430-knot airspeed indicator is seen in figure 28.

Figure 28.—Dial of 430-knot airspeed indicator.

Sources of Error

In the interests of obtaining accurate readings from an airspeed indicator, you should remember that errors are usually the result of installation conditions or of poor adjustments on the instrument. These troubles can be corrected— and must be if the instrument is to do as good a job as it is capable of doing.

No matter how carefully the spot is chosen for placing the pitot-static tube on the fuselage or wing, it is almost impossible to keep it free from all air disturbances set up by the airplane structure. Actually, therefore, the differential pressure developed in the pitot-static system is slightly different than it would be if conditions were theoretically perfect. Allowances must be made when you're reading the indicator for this "installation error."

Errors may also be caused by temperature changes in the instrument, or by imperfect scaling of the indicator dial with respect to the airspeed-differential pressure relationship. Rather simple adjustments can be made in the instru-ment mechanism itself to correct the tendency of the instrument to read fast or slow.

ALTIMETERS

The altimeter in your airplane can do two jobs. It can be used to measure the height of the airplane above some fixed point on the ground—regardless of the airplane's elevation above sea level—or it can measure the airplane's altitude with respect to sea level.

The altimeters installed in military aircraft are of the sensitive type. Their aneroid mechanisms are precision-made, and jeweled bearings are used at all main pivots to reduce friction. A temperature compensator is built into each instrument to help correct errors arising from changes in temperature. And the term "sensitive" is no exaggeration, when you consider that movement in the aneroid of % 6 of an inch will cause the longest dial hand on the altimeter to turn around 35 complete revolutions.

You see the mechanism of a sensitive altimeter in figure 29. The movment of the aneroid diaphragms is linked to a rocking shaft, causing the latter to turn.

DIAPHRAGM

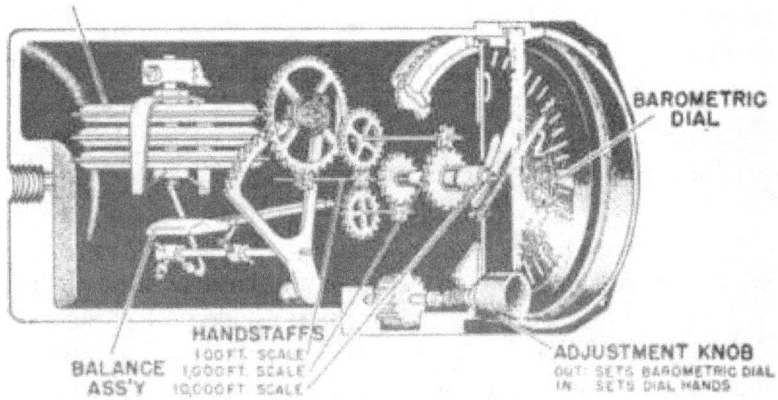

Figure 29.—Mechanism of a sensitive altimeter.

The sector, attached to the rocking shaft, engages a pinion and gear, causing movement of the indicator hands over the face of the dial. There are three hands, or pointers, over the

dial face. The long one—or "minute hand"—makes one complete revolution for each change of 1000 feet. The intermediate or "hour" hand makes 1 revolution for a change of 10,000 feet. Each revolution of the small "second hand" pointer would indicate 100,000 feet except that it's restrained from turning beyond the instrument's limits.

Standard models of altimeters have a calibrated range of from 1000 feet below sea level (-1000 feet) to 35,000 feet above sea level (35,000 feet), but some of the newest models have an extended range and can indicate altitudes up to 50,000 feet above sea level.

Look at the altimeter dial shown in figure 30. The pointers and barometric scale rotate and indicate with reference to the fixed altitude scale which is marked from 0 to 10. A setting knob at the bottom front of the instrument drives a pinion which rotates the pointer mechanism.

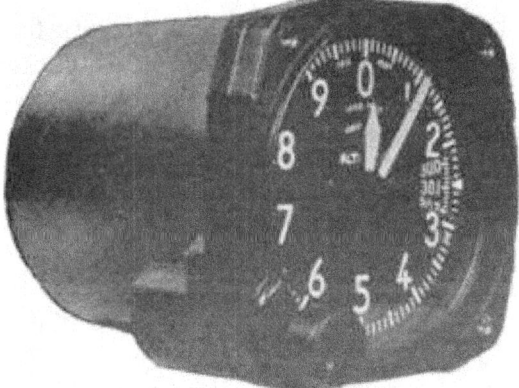

Figure 30.—Dial of a sensitive altimeter.

It is well to note here that some Navy altimeters also have rotating reference markers on the dial, so don't be confused if you see them. The positions of the reference markers are read on the same dial as the pointers. The outer marker makes one revolution for each 1000 feet of altitude change. The inner rotates once for each 10,000 feet.

The standard range for the barometric scale is 28.1 to 31.0 inches Hg. When the range limit of this scale is reached at

either end on some models, a shutter blanks out the barometric dial opening and the barometric pressure is indicated in feet by the position of the reference markers. This can be converted to inches Hg., if necessary, by reference to standard altitude-pressure tables. The fact that there are limits to the range of the barometric scale in no way affects the almost unlimited possibilities of setting barometric pressure by means of the reference markers, if your instrument has them.

An altimeter is enclosed in a two-piece, air-tight case which has a connection at the back for joining it into the static air ijnb of the airplane's pitot-static system. Since the static air line from this system delivers undisturbed air at actual outside atmospheric pressure to the altimeter, the importance of the air-tight case is apparent when the sealed cabins of many modern aircraft are considered.

Setting and Corrections

For short flights or practice in instrument landing, the setting knob is turned so that all the pointers indicate zero. On this setting, the reading on the barometric scale will be the local atmospheric pressure in inches Hg. But if you're setting the altimeter for cross-country flights, turn the knob so that the pointers indicate the known elevation of the field where the takeoff will be made. The barometric scale reading will then be the local pressure reduced to sea level setting.

It is frequently necessary to check the zero-setting error of an altimeter,/ First, you obtain the reading from the station altimeter at the flying field. Then set the pointers of the airplane altimeter to zero, tapping the panel two or three times while making the setting. Next, check that the reading Of the reference markers, if any, is the same as the pressure altitude on the station altimeter—plus or minus a tolerance of 30 feet.

If the error is greater than the permissible tolerance, loosen the small screw to the left of the setting knob. Don't try to pull this screw out. You'll only succeed in doing serious damage if you pull it. When the screw is loose, push

it to the left as far as it will go. Then pull out the setting knob and turn it until the reference markers correspond with the pressure altitude you obtained from the station altimeter. Finally, push the setting knob in, reseat the screw, and tighten it.

Remember that when you're checking and setting altimeter reference markers, the corrections on the scale correction cards of the station and airplane altimeters must be taken into account. The correction cards provide the information needed concerning the scale errors inherent in that particular altimeter mechanism. These errors are given at two temperatures so that you can determine the error at any existing instrument temperature. The scale errors change with instrument temperature because the automatic built-in temperature compensator is not perfect.

The scale correction card is placed in a holder on the instrument panel, as close as possible to the altimeter dial. If it is necessary to replace an altimeter in an airplane, the scale correction card must be removed from the holder and attached to the instrument.

Altimeter Checkup

To check the case for leaks, disconnect the static tube from the instrument and place a short piece of rubber tubing over the vent. Suck on the tube to increase the altitude reading by about 1000 feet. Then pinch the tube shut, and watch the pointer for 1 minute. It shouldn't change position more than 50 feet during that time. If you're sure all line connections are tight but the pointer still changes* position too rapidly, the leak is probably around the cover glass or adjusting-knob shaft, and the instrument should be removed for repair.

RATE OF CLIMB INDICATORS

A rate of climb indicator—or "climb indicator," as it's commonly called—indicates the rate at which an airplane is climbing or descending. It's an instrument of primary importance for flying at night, through fog or clouds, or

whenever the horizon is obscured. Another use is to check on whether the maximum rate of climb is being secured during performance tests or in actual service.

The rate of altitude change, as shown on a climb indicator dial, is positive in a climb and negative in a dive or glide.

Figure 31.—Rate of climb indicator dial.

In other words, the dial pointer, as shown in figure 31, moves in either direction from the zero point, depending on whether the airplane is going up or down. In level flight, of course, the pointer remains at zero.

HOW THEY WORK

The Navy uses more than one type of climb indicator, the fundamental difference being their range. One type measures altitude changes up to 2000 feet per minute; the others have a range up to 6000 feet per minute. The case of a climb indicator is airtight except for a small connection through a restricted passage, or leak, to the static line from the pitot-static tube, as illustrated in figure 32.

In a climb indicator, therefore, you have two chambers— one of them flexible (the capsule), and one inflexible (the instrument case). Both chambers receive air at atmospheric

RESTRICTED PASSAGE

Figure 32.—Operation of a climb indicator,

pressure from the static line. When your aircraft is on the ground, or in level flight, the pressure in each chamber is the same. But suppose you are in a climb. As the aircraft rises, the atmospheric pressure becomes lower and lower. What happens in your instrument ? Just this—

The pressure in the capsule drops immediately because it is connected to the outside air by an unrestricted passage. The pressure in the instrument case starts to drop too, but can't drop as fast as the pressure in the capsule because its outlet to the atmosphere is so tiny. Thus, the somewhat higher pressure in the instrument case tends to collapse the capsule—just as in an altimeter—and the mechanism connecting the capsule to the dial pointer goes into action to indicate a climb.

When you level the airplane off, the pressure in the instrument case gets a chance to "catch up" and equalize the pressure in the capsule. The walls of the capsule return to normal position, and the pointer returns to zero.

In a dive, the pressure conditions are reversed. The capsule pressure immediately becomes greater than the pressure in the rest of the instrument case. Hence the capsule expands and operates the pointer mechanism to show your rate of descent.

AO

Navy climb indicators which have ranges of 2000 feet per minute are installed in such craft as patrol planes and gliders, which operate at low rates of climb and descent. The 6000 feet per minute instruments are used in fighter planes, scouting planes and other types capable of fast climbing and diving.

Don't be concerned if you discover that a climb indicator lags behind your airplane's actual climb or dive. That is part of the nature of the instrument. When the airplane departs from level flight, it may be several seconds before the pointer indicates the fact. Also when you level off after a climb or glide, the instrument may continue to indicate assent or descent for as much as 12 seconds.

Errors and Corrections

Temperature changes have a tendency to cause variations in the flow of air through the small restricted leak passage of the instrument case. Various manufacturers of climb indicators compensate their instruments for this in different ways. The main thing to remember is that all Navy instruments are equipped with temperature compensating features.

A leak in the instrument case of a climb indicator will obviously mean incorrect readings by causing too rapid pressure changes in the case. The pointer will, therefore, indicate less than the actual rate of climb. It's your job to locate and correct these leaks.

A break in the leak passage (or "diffuser") assembly will have a similar effect on instrument operation, as such a break , also allows instrument-case pressure to change too quickly. Don't try to repair a broken diffuser. Replace the broken assembly with a new one.

If the pointer fails to respond, the condition is usually caused by an obstruction in the static line. To correct this, disconnect all instruments which may be connected to the static line, open the line's drain plug, and blow the line clear. Oscillation of the pointer is usually the result of a leak in the static line. To check on this, disconnect all instruments from the line and test them, as well as the static line, to find the leak.

The zebo adjusting knob on the front of the case can be used while the airplane is on the ground to return the pointer to zero. Tap the instrument lightly to remove friction effects while you're making this adjustment

QUIZ

L The pitot-static tube operates all but the _

a. Altimeter

b. Rate-of-climb indicator.

c. Vacuum gage.

d. Airspeed indicator.

2. The static tube furnishes air.

3. Remember to so as to remove friction effects during tests of either static or pitot lines.

4. Errors in airspeed indicators are usually the result of

conditions or of on the instrument.

5. The altimeter can be used not only to measure altitude above sea level, but also to

The inner marker of a Navy altimeter rotates once for each

feet

a. 1000.

b. 10,000.

c. 100.

d. 10.

7. For short flights or practice instrument landings, the altimeter setting knob is turned so that all pointers are ,,

8. The provide the information needed concerning the scale errors inherent in a particular altimeter mechanism.

9. The indicates the rate at which the aircraft is descending or climbing.

10. The rate-of-climb indicator may lag as much as —

a. One second.

b. One minute.

c. Twelve seconds.

d. Thirty seconds.

11. The of a pitot-static tube receives impact pressure of the atrstream.

12. When removing a pitot-static tube, first disconnect the two .

13. To check static lines for leaks,

a. Apply 1000 pounds of pressure.

b. Apply vacuum slowly.

c. Blow through them gently.

d. Run water through them.

14. An airspeed indicator is fed with how many different types of pressures during flight?

15. Standard models of altimeters have a calibrated range of from .

a. 1000 feet above sea level to 35,000 feet above sea level.

b. 1000 feet below sea level to 35,000 feet above sea level.

c. Sea level to 35,000 feet above sea level.

d. 5000 feet below sea level to 35,000 feet above sea level.

16. If an altimeter has a leak around the cover glass or adjusting* knob shaft it should be .

17. The rate-of-climb indicator is ~.

a. Completely airtight.

b. Not airtight

c. Airtight except for a small connection.

d. None of the above.

18. Rate-of-climb indicators for Navy fighter aircraft have a range of feet per minute.

19. To return the rate-of-climb pointer to zero when the aircraft is on the ground,

20. Pitot-static tubes are made of or chrome-plated

21. When an airplane is not flying, its airspeed tube is

22. The standard range of the altimeter barometric scale is inches of mercury.

23. When checking the zero-setting of an altimeter, first .

24. The pointer of a rate-of-climb indicator moves from the zero point

a* Up.

b. To the left

C. To the right

d. In either direction.

25. There are chambers in the rate-of-climb indicator.

a. Two*

b. Three.

c. Pour.

d. Many.

26. Oscillation of the rate-of-climb pointer is usually a result of
CHAPTER

* THERMOMETERS THE BI-METAL THERMOMETER

It's important for a pilot to know the temperature of the air surrounding his airplane. If, for instance, the moisture content of the air is high, there is danger of ice formation on airplane wings at freezing temperature. The outside air temperature also must be known in order to make corrections of altimeter readings and thus to find the airplane's correct altitude.

One type of outside air thermometer used on Naval aircraft uses two strips of different metals which expand and contract at different rates when heated or cooled the same amount.

Nearly all solid materials expand when heated. If you measure the amount of expansion that takes place for a given increase in temperature in various solids—such as brass, or iron— you can make a list of the amounts and compare them to one another. Scientists have already prepared such a list, and have assigned values to the expansion rates of almost all materials. These values are represented by numbers. The number used for any specific material is called its coefficient of expansion. Thus iron and brass have coefficients of expansion which tell you that brass expands more than iron when heated the same amount.

Now, suppose you have two strips of metal, one brass and the other iron, and fasten them together as in figure 33A. At average room temperature, it is assumed, the lengths of the two strips are the same. As the strips are heated— or as the temperature of the room increases—the length of the brass strip increases more than the length of the iron strip, and the bi-metal strip curves as shown in figure 33B.

274033'

Figure 33 —(A) Bi-metal strip at room temperature. (B) After heating. (C) After cooling.

When the temperature falls, the brass strip contracts more ithan the iron, and the curvature will be reversed as you see in figure 33C.

• By fastening an indicator pointer to one end of the bimetal strip and attaching the other end of the strip rigidly to an instrument case, you have made yourself a thermometer of the bi-metal type. Place a dial marked in degrees of temperature behind the indicator pointer and you can observe its readings whenever you wish.

Many of the bi-metal outside air thermometers used on Naval aircraft are made just that way. They have a range of from —40° C. to +40 ° C, and are mounted outside the cockpit so that the dial is clearly visible to the pilot. The cases of these thermometers are designed so as to offer the least possible air flow resistance, and should be installed with face parallel to the fore-and-aft axis of the airplane at a point relatively free of vibration. Since the outsides of the cases are chromium plated to reflect heat from other sources i than the air, they should not be marred or painted.

The bi-metal strip in some thermometers is in a spiral shape, as shown in figure 34. The pointer shaft is attached directly to the bi-metal strip at the center of the spiral, and the unit is enclosed in a light case—rain-tight except for a drain hole in the bottom.

If a check of the instrument shows that the readings are incorrect, you can reset it by removing the glass at the front of the dial, seen in figure 35. Remove the pointer by

Figure 34. Bi-meta! spiral in outside air thermometer*

Figure 35.—Bi-metal thermometer dial.

means of a pointer jack, or by slipping two small screwdrivers under the pointer on opposite sides of the shaft and lifting the pointer off. Then replace the pointer in the correct position. Don't disturb the bi-metal strip. Its proper length is fixed at the factory and it is therefore permanently adjusted.

VAPOR PRESSURE THERMOMETERS

Aircraft are equipped with a number of other types of thermometers besides those which measure and record the temperature of outside air. Some of these types of thermometers can be used for any one of several different tem-perature-measuring jobs. One such many-purpose thermometer is the vapor pressure type. It can be used to indicate the temperature of air inside or outside the cockpit, of cooling liquid in the radiator of a liquid-cooled engine, of lubricating oil, or of the carburetor mixture.

Vapor pressure thermometers were formerly used more frequently than they are today, but you'll still find them on some aircraft. They consist of three main parts—the bulb, the capillary tube, and the indicator. The range of an individual instrument depends upon the use to which it is put. Some capillary thermometers have a range of —10° to +20° Centigrade, while others range from +32° to 212° Fahrenheit.

The bulb in these instruments is a hollow metal cylinder, tightly sealed and containing a volatile liquid such as methyl chloride. The capillary tube is a copper pipe of very small diameter, protected by braided copper wire armor. The indicator is much like some of the pressure gages you read about in an earlier chapter, operating by means of a Bourdon

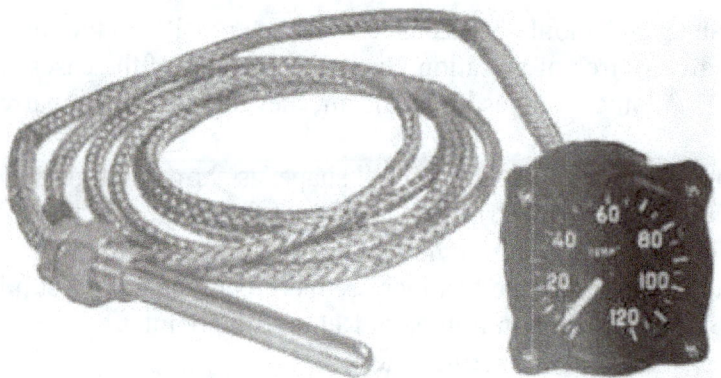

Figure 36.—Vapor pressure thermometer, showing bulb, capillary and indicator.

tube mechanism in the instrument case. A bulb, capillary, and indicator can be seen in figure 36.

The bulb is installed at the point where the temperature is to be measured. As the temperature of the bulb increases, the volatile liquid inside it is changed to a gas, and the pressure inside the thermometer system increases. The pressure is transmitted to the Bourdon tube of the indicator through the capillary tube. The Bourdon tube thereupon tends to straighten out, and its movement is indicated by the dial pointer, to which it is linked.

There are two facts that should never be forgotten when you're working with a vapor pressure thermometer—first, that the bulb, capillary and indicator can't be separated without causing serious damage to the instrument; second, that the capillary must not be cut, broken, dented, mashed, stretched or pulled tight.

Friction tape should be shellacked on the capillary tube at all points of contact with other surfaces. This delicate tube should never come in contact with exhaust stacks or hot parts of the engine. The bulb is built to stand the heat. The capillary is less hardy. When it's necessary to replace an instrument, one with a longer capillary may be used if one with the correct length isn't available. Excess length may be taken up by coiling the tube carefully and taping it to some structural part of the engine nacelle or fuselage. • Be sure that every thermometer dial in aircraft you work with is properly labeled to show what temperature it measures. Labeling can be done by attaching name plates, or by marking the indicator with aircraft enamel. Important ? Certainly. Several thermometer dials in a row can look very much alike if the purpose of each isn't clearly marked.

ELECTRICAL RESISTANCE THERMOMETERS

Electrical resistance thermometers find practically the same uses in aircraft as do vapor pressure instruments, and are employed to a greater extent. The main parts of the electrical resistance thermometer are the indicating instrument, the TEMPERATURE-SENSITIVE ELEMENT (or bulb), and

Figure 37.—Indicator of an electrical resistance thermometer.

the connecting wires leading from the bulb and equipped with plug connectors.

Oil temperature thermometers of the electrical resistance type have ranges of from -10° C. to +120° C. to -70° C. to 150° C; some carburetor air and mixture thermometers have a range of —50° C. to +50° C, as do many free air thermometers. The oil, coolant, and carburetor thermometers are made in dual form for use with multiengine airplanes, as well as in single form. Indicator dials of oil, carburetor, and free air thermometers are much like that in figure 37. The indicators are self-compensated for changes in cockpit temperature.

How do electrical resistance thermometers work? They're based on the scientific fact that most metals change their electrical resistance with changes in temperature. In almost every case, the electrical resistance of a metal increases as the temperature rises. The resistance of some metals, of course, increases more than the resistance of others with a given rise in temperature. If a metallic resistor with a high temperature-resistance coefficient—that is, a high rate of resistance rise for a given increase in temperature—is sub-

jected to the temperature you want to measure, and a resistance indicator is connected to it, you have an electrical resistance thermometer.

The heat-sensitive resistor is the main element in the bulb. It is made so that it has a definite resistance for each temperature value in its working range. The indicator is a resistance-measuring instrument with its dial calibrated in degrees of temperature instead of ohms. It actually is a modified form of the resistance-measuring apparatus known as the Wheatstone bridge.

You will remember how the Wheatstone bridge works if you studied physics in school. Its principle is that of balancing one unknown resistor against other known resistances. The simplest form of bridge is shown in figure 38.

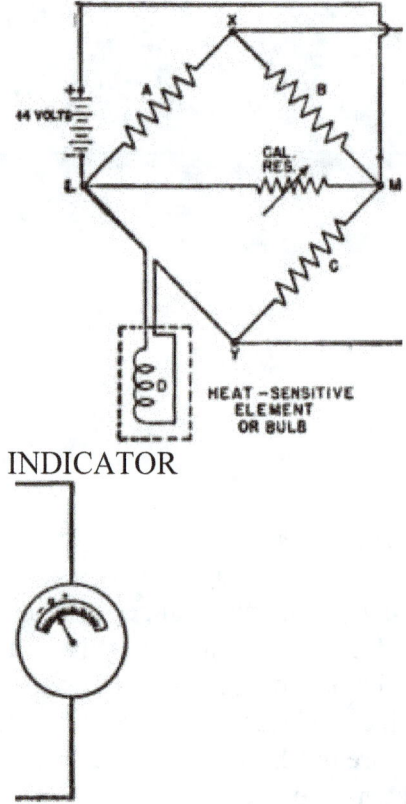

INDICATOR

MEAT -SENSITIVE ELEMENT OR BULB

If you were constructing such a bridge in a laboratory, you would connect three equal known resistances (A, B, and C) and the unknown resistor (D) to form a diamond; and you would attach a galvanometer across the diamond at points (X) and (Y).

The unknown resistor, of course, corresponds to the bulb of your airplane thermometer. If the temperature of this unknown resistor is changed, its resistance will change. But

suppose the temperature is such as to make its resistance in balance with that of A, B, and C. In that case, there will be no flow of current through the galvanometer, and its reading will be zero. But if the temperature of the unknown resistor is raised or lowered, its resistance will also increase or decrease. As soon as a change happens, the bridge is unbalanced, and current flows through the galvanometer in one direction or the other.

Now, simply by calibrating the galvanometer dial in degrees of temperature instead of in amperes, you convert it into a temperature-measuring instrument. And, basically, that's how an electrical resistance thermometer is made. Refinements have been added to the construction of airplane thermometers to make certain that they read correctly even though the battery-voltage may vary a little—but the fundamental principle of operation is the same.

The temperature-sensitive resistor element of the bulb is a winding of kovar or nickel-manganese wire on suitable insulating material. The resistor is protected by a closed-end monel metal tube, attached to a threaded plug with a hexagon head, as you see in figure 39. The two ends of the nickel

Figure 39.—Resistance thermometer bulb assemblies.

winding are brazed or welded to an electrical receptacle, made to receive the two prongs of the connecting plug.

When you are unpacking a resistance thermometer and preparing to install it in an airplane, inspect the indicator carefully to make sure the glass is not broken, and that the pointer swings freely when you rotate the instrument gently.

To learn how electrical resistance thermometers should be wired when installed, be sure to consult the latest AN publications.

Before turning on the battery current, be certain the indicator dial pointer is at the correct position on the scale. On types of instruments which read zero with current off, you can adjust the zero setting by turning a small screw on the front of the instrument with a screwdriver. Never make this adjustment when the current is on. Unless the bulb is damaged by excessive heat it will give accurate service indefinitely. If it is damaged by too much heat—as, for instance, more than 300° C. for a long period—it should be replaced and the old one sent to overhaul. When a thermometer won't work, check carefully for loose wiring connections before blaming it on the bulb.

THERMOCOUPLE THERMOMETERS

The heat generated by an airplane engine is a highly reliable indication of its overall operation. Naturally the pilot wants all the information he can get on the subject, for a properly running engine is a matter of great concern to him. The temperature of an air-cooled radial engine is gaged by a thermometer which has its heat-sensitive element attached to some point on one of the radial cylinders, or the exhaust temperature of the jet aircraft is measured by attaching thermocouples to the tailcone. Thermometers of the thermocouple type are generally used for measuring engine or tailcone temperatures, as a thermocouple is capable of measuring high temperatures.

What does a thermocouple circuit consist of ? It's an electrical circuit formed of two wires of different metals. Such a circuit has two junctions, of course. If one of the junc-

rjgure **u,— i nermocoupie inermomeTer maicaror.

tions is heated to a higher temperature than the other, an electromotive force is produced in the circuit. By including a galvanometer in the circuit this electromotive force can be measured. The hotter you heat the high-temperature junction, the greater the electromotive force. If the galvanometer dial is calibrated in degrees of temperature, therefore, it becomes a thermometer.

The thermocouple thermometers used to indicate engine temperatures in Navy airplanes consist of a galvanometer-type indicator, a thermocouple, and thermocouple leads. Many of the thermocouples and leads are made of copper and constantan—the latter being an alloy of copper and nickel. The Navy uses copper-constantan, iron-constantan and chromel-alumel. Iron-constantan is used mostly in radial aircraft; chromel-alumel is the only type used for jet aircraft. Figure 40 shows an indicator of a type used by the Navy.

The "hot" junction of the thermocouple varies in shape depending on its application. A common type is a copper gasket which replaces the standard spark plug gasket on the cylinder chosen for temperature measurement.

The "cold" junction of the thermocouple circuit is inside the indicator instrument case. Since the electromotive force set up in the circuit varies with the difference in temperature between the "hot" and "cold" junction, however, it's obviously necessary to compensate the indicator mechanism for changes in cockpit temperature which affect the "cold" junction. This is done by means of a bi-metallic spring—much like the one in a bi-metal air thermometer connected to the indicator mechanism.

When you disconnect the thermocouple lead from the indicator, the latter will indicate the temperature of the cockpit. Why ? Because, even though the thermocouple is not working, the bi-metallic compensator spring keeps on functioning as a thermometer.

If an old instrument is to be removed and a new one installed to replace it, you should keep some special points in mind. For instance, you'll find that the binding posts of a new thermocouple indicator are short-circuited by a short piece of copper wire to close the electrical circuit for protection during shipment. This wire should be removed and discarded at the time of installation. Also, the magnetic shield on the instrument must not be removed or the thermocouple indicator will foul the readings of magnetic compasses in the airplane.

Be positive that all connections are clean and tight when you're making an installation. Otherwise, resistance will be introduced and they'll make the readings incorrect. After you have secured the thermocouple gasket in position under the spark plug and made connections to the leads, anchor the connectors with insulation tape or a small clamp to keep them from breaking as a result of vibration.

The connectors, as you may notice by looking at the wiring diagram in figure 41, are bare, and must be taped to prevent short circuiting through contact with the engine.

Thermocouple leads must be of the same material as the thermocouple. The lead resistance (including the thermocouple) must correspond to the ohms resistance for which the indicator is calibrated. Thermocouple leads differ as

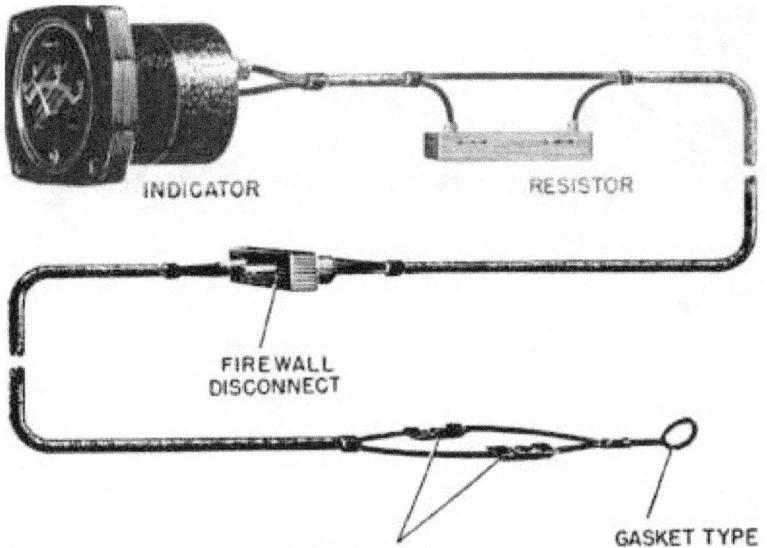

INDICATOR　　　　RESISTOR

FIREWALL
DISCONNECT

GASKET TYPE

TERMINALS THERMOCOUPLE Figure 41.—Wiring diagram for thermocouple thermometer installation.

to type, length, and electrical resistance values. The Erection and Maintenance Instructions should be consulted for the particular type used.

The thermocouple lead should be attached to the thermocouple at one end and to the indicator at the other end. The indicator end of the lead has two eye terminals with holes of unlike diameter. The thermocouple end of the lead has one eye terminal, and another terminal with a nut mounted on it. The plus and minus studs of the indicator are of different diameters. They're made that way so that connections to the lead can't be made improperly.

Here's a warning. The thermocouple leads must not be lengthened or shortened. If it is necessary to replace the leads, be sure they are from the same size stock and the same length as the original. They are of a definite, calculated resistance, and the calibration of the indicator depends upon the lead resistance remaining unchanged. Whenever the thermocouple leads are to be replaced, refer to the applicable AN publication for complete data on adjustment procedure.

When you check the indicator for its zero position, open the circuit of the thermocouple, as by disconnecting one of the eye terminals from the stud on the back of the indicator. The pointer should then read the temperature in the cockpit, as was pointed out previously. This reading can be checked by placing an ordinary glass-tube mercury thermometer close to the indicator—allowing time, of course, for the mercury thermometer to adjust itself to cockpit temperature. The zero position of the indicator can then be adjusted to agree

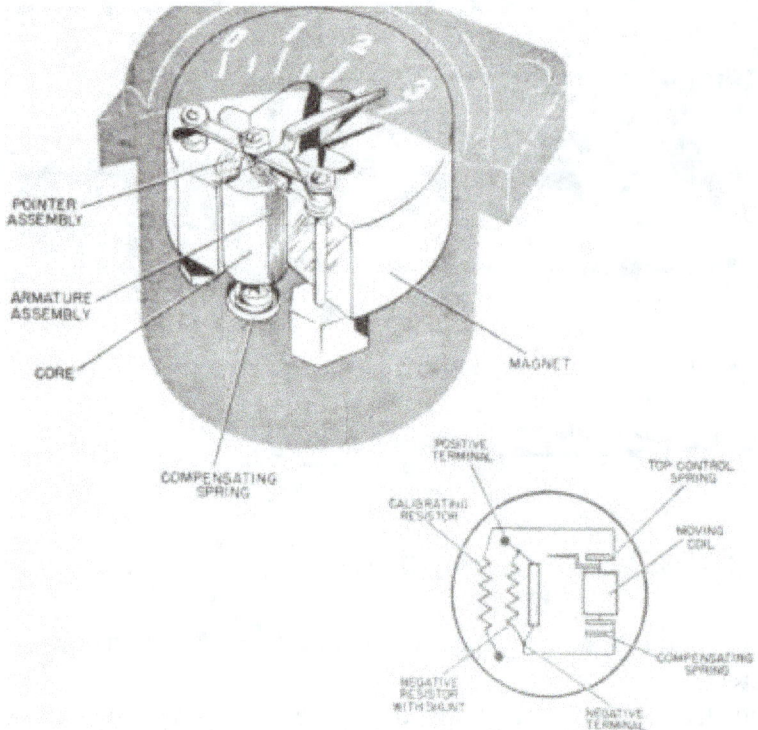

Figure 42.—Internal wiring of a thermocouple thermometer.

with the mercury thermometer reading by turning the adjusting screw that you find on the indicator face.

The indicator is fundamentally a millivoltmeter of the permanent magnet moving coil type, which consists essen-

tially of a moving coil mounted and pivoted in the field of a permanent magnet. The two leads from the thermocouple are connected to the ends of the moving coil and a pointer attached to the coil moves across the scale calibrated in degrees Centigrade. The moving coil will rotate as voltage is applied to it, the deflection being proportion to the voltage. Since the voltage is a function of the temperature difference between the indicator and thermocouple temperatures, the pointer will indicate directly in terms of the cylinder head temperature. As stated before, the indicator is compensated by means of a bimetallic spring for variations in cockpit temperature so that the indicator will indicate the true temperature of the cylinder head and not just the difference between the indicator and thermocouple temperature. The internal wiring is shown in figure 42.

The operation of the indicator is entirely automatic and requires no attention from the pilot other than observing temperature indication during flight.

QUIZ

1. A bi metal thermometer uses two strips of different
2. Brass expands than iron for the same amount of heat.
3. The cases of thermometers are chromium plated to
4. Vapor pressure thermometers have three main parts: The , the , and the
5. All thermometers should be to show what temperature is measured.
6. The main parts of the electrical resistance thermometer are the , the , and the connecting wires.
7. The electrical resistance thermometer operates on a modified bridge.
8. When a thermometer won't work, first check for .

9. A thermocouple is an electrical circuit formed of .

10. The ,, hot" junction of a thermocouple is a

11. The outsides of Navy bi-metal thermometers are

a. Corrugated.

b. Painted black.

c. Gold plated.

d. Chrome plated.

12. The capillary tube of a vapor pressure thermometer Is made of

13. Oil temperature thermometers of the electrical resistance type have a range of to centigrade.

a. —50 degrees to 120 degrees.

b. —10 degrees to 120 degrees.

c. 0 degrees to 212 degrees.

d. —10 degrees to 220 degrees.

14. The principle of the Wheatstone's bridge Is the balancing of one unknown against other

15. Navy thermocouple leads are made of and .

16. When you disconnect the thermocouple lead from the indicator, the indicator will show the temperature of the

17. The thermocouple indicator Is fundamentally a

18. Navy bi-metal thermometers used on aircraft have a range from degrees C to degrees centigrade.

19. Vapor pressure thermometers consist of three main parts. The is not one of them.

a. Bi-metal strip.

b. Bulb.

c. Capillary tube.

d. Indicator.

20. The thermometer uses a bourdon tube in the indicator.

21. The main element in the bulb of an electrical resistance thermometer is the •

22. When connecting a thermocouple thermometer, be sure all connec-nections are and

CHAPTER

TACHOMETERS ELECTRICAL TACHOMETERS

Why do airplanes have tachometers? Primarily to let pilots know how fast the propellors are turning. Pilots have to keep tabs on propeller speeds for many reasons. Nowadays, for example, combat service reciprocating engine airplanes are almost all equipped with either constant-speed or adjustable-pitch propellers. "When a pilot flies an airplane which has constant-speed propellers, he uses the tachometers to check on the operation of propellor governors and controls during takeoff, and during variable altitude and density conditions. With adjustable-pitch propellers, he depends on tachometers as a check when he shifts the blades from low to high—or high to low—pitch.

Actually a tachometer measures the number of revolutions per minute made by the engine crank-shaft. If the pror peller is connected directly to the crankshaft, the r.p.m. of both will be the same. If it's geared to the crankshaft, the propeller r.p.m. will still have a definite proportion to the crankshaft speed, even though the propeller is turning at a different rate of speed.

The tachometers used by the Navy consist of two units— an indicator and a generator, connected by an electrical cable. The generator unit is driven by the engine to which it is attached, and generates an alternating current which is transmitted to the indicator by the cable. The indicator consists primarily of a motor which rotates an assembly of magnets. The speed at which the moving parts of the motor turn depends directly on how fast the generator is turned by the airplane engine.

274033 •—54 7

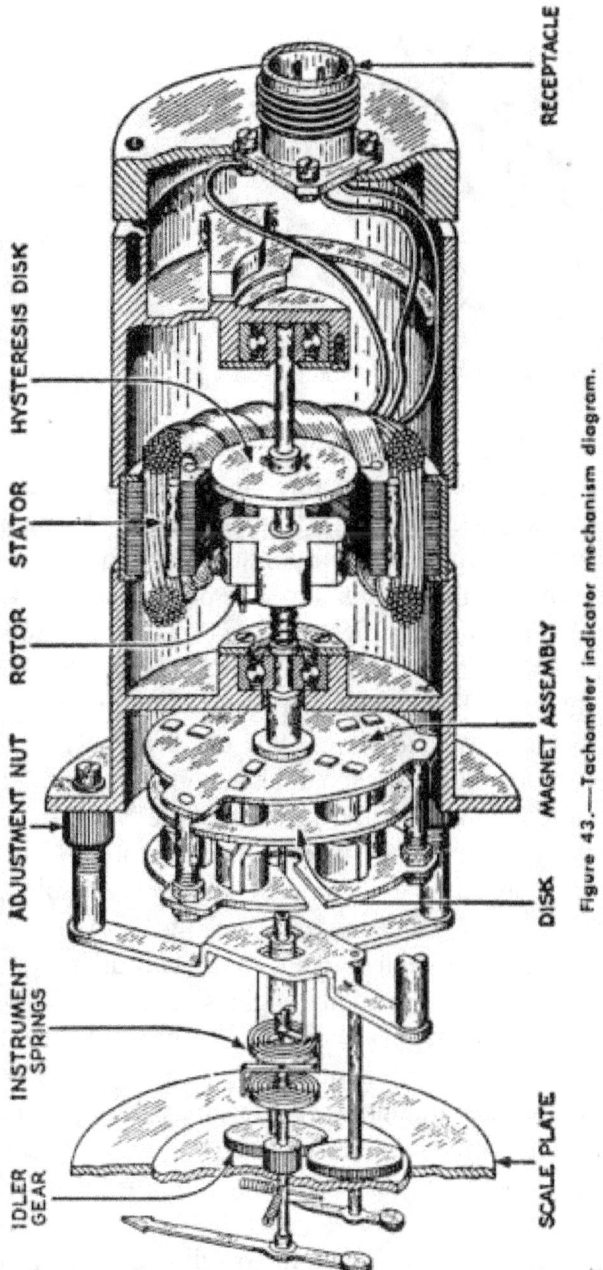

RECEPTACLE

HYSTERESIS DISK

STATOR

ROTOR

ADJUSTMENT NUT

INSTRUMENT SPRINGS

IDLER GEAR

MAGNET ASSEMBLY

DISK

SCALE PLATE

Figure 43.—Tachometer indicator mechanism diagram.

A metal disk—or drum—is located beside the motor-turned magnet assembly in the indicator. The, disk is mounted axially on a shaft separate from the motor shaft. You can familiarize yourself with a typical indicator mechanism by looking at figure 43. As the motor turns the magnets, the disk is affected by the spinning magnetic fields and tends to rotate with the magnets.

The disk is restrained from revolving by hair-springs attached to its shaft, however, and can turn only part of one revolution. The angle to which it does turn depends on the speed of the revolving magnets. The disk shaft is geared, or otherwise connected, to the indicator pointers, as shown in figure 43. The indicator-dial reading shows the speed at which the airplane engine is turning, as it is calibrated in terms of r.p.m.

Figure 44.—Tachometer indicator dial.

The range of an indicator is from 0 to 3500 r.p.m. or from 0 to 4500 r.p.m. Dials of Navy types are calibrated in 10 major steps, each representing 100 r.p.m. as you can see in figure 44. Each of these steps is divided into smaller units— of 10 to 20 r.p.m.—by intermediate markings.

The indicator dial is also fitted with a smaller pointer which indicates r.p.m. in units of 1000. Thus when the r.p.m. goes higher than 1000, the large pointer simply starts over again from zero and the small pointer indicates that 1000, 2000 or 3000 r.p.m. are to be added to the reading. The small pointer is geared to the large one in a ratio 1 to 10, as you can observe by referring again to figure 43. The numerals, pointers, and major marking lines on the dial are coated with luminous material so as to be readable under night flying conditions

If an airplane has two or more engines, a separate tachometer can be installed for each, or dual indicators can be used. It's a good thing to know, too, that the Navy's electrical tachometer indicators and generators—even though made by various manufacturers—are interchangeable.

Tachometer generator units are small and compact, and are available in two types, known as the stud type and the screw type. These names refer to the kind of mounting used in attaching the generator to the engine. Engines having a threaded tachometer outlet require the use of a screw type tachometer generator. The outlets of other engines have square, pads for mounting the generator, and take the stud type. Both kinds of generators are seen in figure 45, and either can be used with any current model indicator instrument.

Generators are made for universal mounting. That is, they can be attached to an engine so as to face the direction you find most convenient during installation. This feature is important as there frequently are obstacles that would interfere with installation if the generator could be attached in only one position.

The stud-type generator is mounted on the engine by alining the key on the generator shaft with the keyway in the engine sh#ft and then fastening the generator securely in the engine. The screw-type generator should be alined in a similar way and then screwed on the outlet. Be sure to clean the outlet threads. After such a generator is mounted, it may be necessary to change its position so the electrical cable connector will be in the place you wish. Do this by removing the cover screws and turning the covers 90° or 180° with respect to each other, then replacing the screws.

Figure 45.—Tachometer generators—right, stud type; left, screw type.

When you're mounting a screw-type generator, it's always a good idea to provide some additional support for it. A bracket fastened to the engine or airplane framework can be rigged as such a support to keep the generator from loosening on its mounting when subjected to vibration.

The tachometer indicator should be mounted on the instrument panel in the engine instrument group. The zero mark on the dial is at the topside. The indicator is connected to the generator by a two-wire or three-wire cable, equipped at each end with mating plugs which fit into receptacles on both the generator and indicator. The third connection is made via grounding. If the indicator rotates in the wrong direction during operation, you can correct the situation by interchanging the lead wires from the generator.

Most Navy tachometer indicators have a zero setting attachment, which makes it possible to correct a small amount of pointer error. If you have to make such an adjustment, first remove the four screws which hold the cover flange to the frame of the instrument, and then lift off this cover flange.. Then take a small open-end wrench to reach and loosen the clamping screw on the zero set. You can now move the lever to make the necessary correction. Finally, tighten the clamping screw and replace the cover.

If an indicator reads too high over its full scale, it is necessary to decrease the strength of its magnets. You can do this by bringing an alternating-current "knockdown" coil close to the indicator within the plane of the magnet. But be careful when doing this job. If magnet strength is reduced too much, readjustment of the magnet assemblies will

have to Iks made. If the indicator originally reads too low, adjustment can be made by loosening the nuts holding the magnet assemblies in place and turning the adjusting nuts counterclockwise. This brings the magnets closer to the disk, and increases their effect on it.

SYNCHROSCOPES

Any farmer can tell you that two horses hitched as a team will pull a wagon most efficiently if each horse does its full share of the work. The same thing is true when airplane engines are concerned. On a multiengine airplane, all power plants must pull evenly for maximum combined efficiency. If they don't, power is lost and vibration and throbbing will tend to shake your airplane loose at the seams.

The synchoscope is an instrument which tells whether or not a pair of engines are synchronized—that is, whether they are turning the same number of r.p.m. The instrument consists of a small electric motor, which receives electrical energy from the tachometer generators of both engines that are to be synchronized. The synchroscope is made so that energy from the faster-running engine's generator controls the direction in which the synchroscope motor rotates.

How does this work out? If both engines—and thus both generators—are running at exactly the same speed, the synchroscope motor doesn't turn. If, however, one engine is turning more rapidly than the other, its generator will turn the synchroscope motor, say, to the right. If the other engine speed gets faster than the first, then its generator will have the upper hand. The synchroscope motor will then reverse itself and rotate to the left.

The motor of the synchroscope is connected by means of a shaft to a double-ended pointer on the dial of the instrument, as shown in figure 46.

As you can see from the following description and from the illustration, it is necessary to designate one engine of the airplane as the master engine if the synchroscope indicators are to have any meaning. The dial readings—with

Figure 46.—Synchroscope dial.

leftward rotation of the pointer indicating "slow" and right-ward rotation indicating "fast"—would then refer to the operation of the SECOND engine in relation to the speed of the master engine.

After one of the engines has been selected as the master or "first" engine, the wiring diagram shown in figure 47 should be followed in connecting up the synchroscope to be sure that the pointer rotates in the proper direction. If you

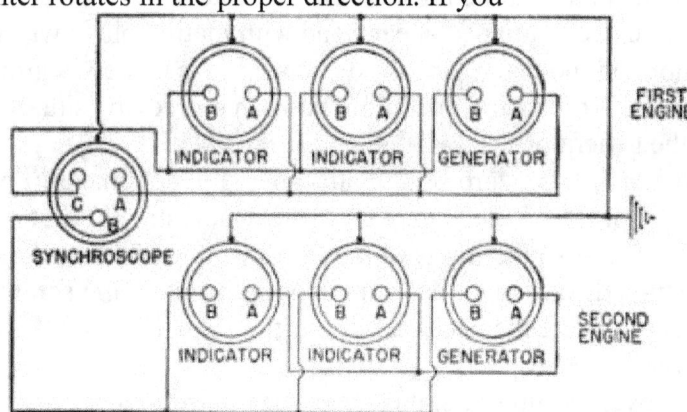

Figure 47.—Synchroscope and tachometer wiring diagram.

find that you've made a mistake and the pointer rotates the wrong way, you can correct the error by interchanging the lines A and C on the synchroscope connector plug. Notice that the synchroscope is wired in the same circuit with the tachometer indicators, with its energy being supplied by the same generators.

For airplanes having more than two engines, additional synchroscopes are used. One engine is picked as the master,
BRUSH

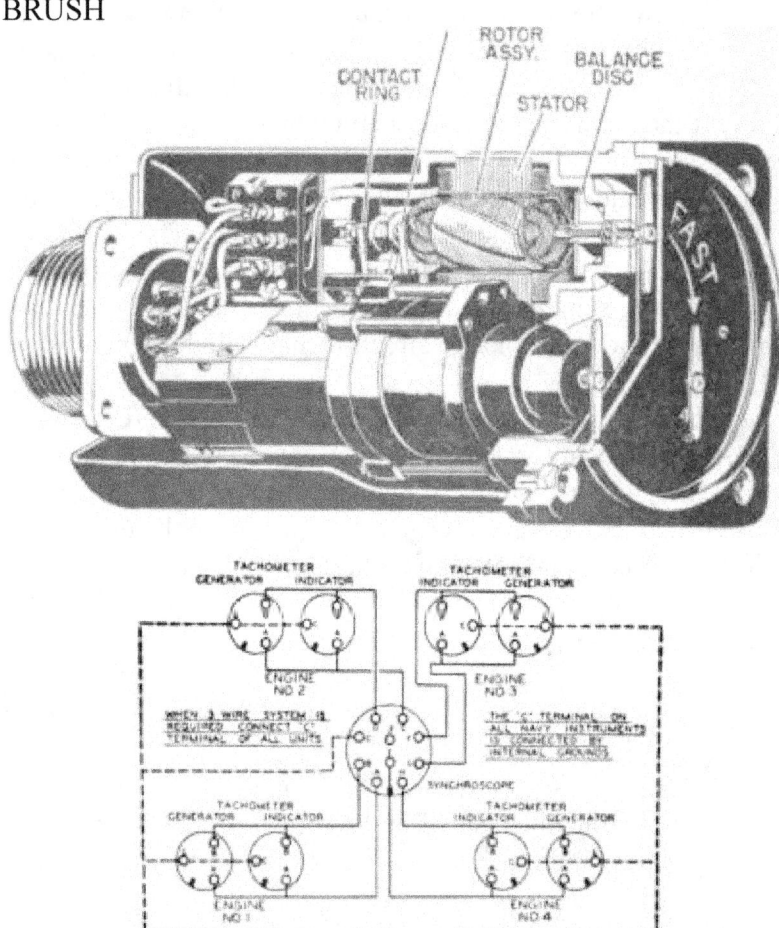

Figure 48.—Four-engine synchroscope.

and synchroscopes are connected between its tachometer and those of each of the other individual engines. On a complete synchroscope installation of this kind, there will, of course, be one less instrument than there are engines, as the master engine is common to all the pairs.

Four-engine Synchroscope

The four-engine synchroscope is a special adaptation of the synchroscope intended for use on a four-engine aircraft to indicate when the engines are synchronized. This instrument is actually three individual synchroscopes placed in one case.:

The rotor of each is connected electrically to the tachometer generator of the engine designated as a master while each stator is connected to one of the other engine tachometers. There are three hands, and each hand indicates the relative speed of the number two, three, or four engine as shown in figure 48.

The separate hands revolve clockwise when their respective engine is running faster than the master and counterclockwise when it is running slower. An excessive speed difference does not cause the hand to revolve faster as might be supposed, but does cause a rocking motion of the hand. Rotation of the hand begins as the difference reaches about 350 r.p.m. and, as the engines approach synchronism, the hand revolves at a rate proportional to the speed difference.

Dual Tachometers

With the increasing necessity of more and more instruments for efficient flight, the combination of several instruments in one definitely becomes an added attraction. In line with this, the dual tachometer indicator has been developed for use in multiengine and turbojet aircraft.

This unit consists of two synchronous-motor-magnetic-drag tachometer indicators housed in a single case. The indicators show simultaneously on a single dial the speeds of rotation of the engine crankshafts in revolutions per minute.

One tachometer is used for eacli two engines or turbojets of the aircraft.

Dual tachometers have also been placed in the same case with a synchroscope for various purposes. One of these is the helicopter tachometer with synchroscope, figure 49, an instrument used to show simultaneously the speed of rotation

Figure 49.—Helicopter tachometer with synchroscope.

of the engine crankshaft, the speed of rotation of the rotor shaft, and the slippage of the rotor due to malfunctioning of the clutch or excessive speed of the rotor when the clutch is disengaged in flight. The speed of the rotor shaft and the engine shaft are both indicated by a regular dual tachometer while the slippage is indicated on a synchroscope.

QUIZ

1. A tachometer measures the number of revolutions per minute made by the

2, Navy tachometers consist of two units: and

:\. The two units of Navy tachometers are connected by a(n)

4. Tachometer generators are made for mounting.

5. The synchroscope is an instrument which indicates whether two engines are

6. When a synchroscope is installed, one engine is designated as

7. The synchroscope receives its energy from the

8. On a complete synchroscope installation there will be i-synchroscopes as engines.

a. The same number of. h. One fewer.

c. One more.

d. Half as many.

9. The four-engine synchroscope is actually how many instruments in one case?

10. The helicopter tachometer synchroscope is used to indicate the of the rotor.

11. Navy tachometers made by different manufacturers are

12. Generator units are available in two types: and type.

13. If a tachometer indicator indicates too high for its full scale, it is necessary to

14. If both engines of an aircraft are running at the same speed, the synchroscope .

15. The dial readings of a synchroscope refer to one engine in relation to its engine.

16. Multiengine and turbojet aircraft use tachometer indicators.

17. Navy tachometers range from 0 to ———— r.p.m.

18. A screw-type generator should always have . to keep it from vibrating loose.

19. Three individual synchroscopes in one case are used in aircraft.

20. The helicopter tachometer with synchroscope is a special adaptation of the tachometer.

CHAPTER

8

FUEL GAGES AND FLOWMETERS

D-C SELSYN FUEL GAGES

Fuel-quantity indicating equipment of the d-c selsyn type consists of an indicator on the airplane instrument panel and a transmitter mounted in each of the fuel tanks. The transmitters and indicators of these gages are identical in principle with those of the d-c selsyn position indicators you read about in chapter 3.

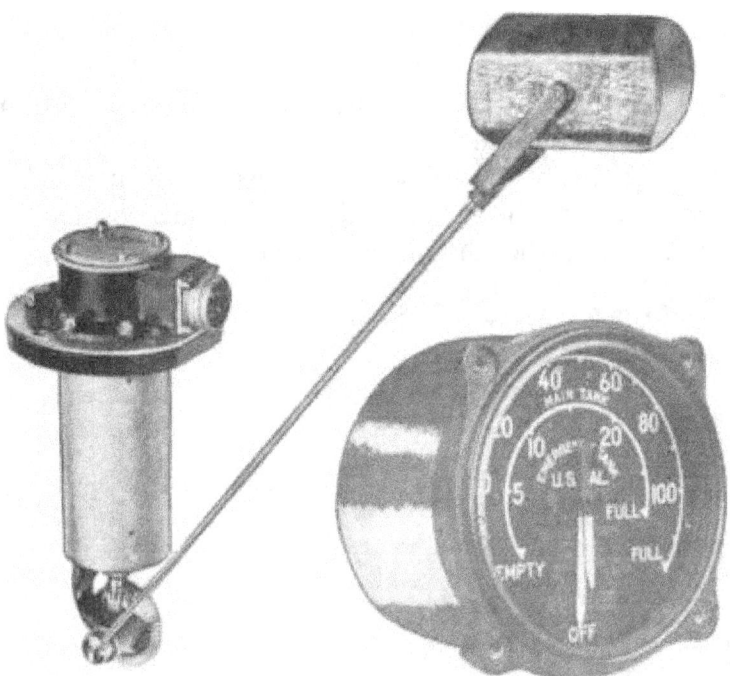

Figure 50.—Dial (right) and transmitter unit (left) of d-c selsyn fuel quantity

A typical transmitter unit and indicator dial are shown in figure 50. As the fuel quantity changes inside the tank, the position of the transmitter float is also shifted. The motion of the float is transmitted through a pair of gears to a U-shaped magnet. A diamond-shaped magnet inside the U-magnet is attached to the shaft of the transmitter resistance element, and follows the motion of the U-shaped magnet. This motion rotates the transmitter element shaft, and an indication of the rotation is carried electrically to the indicator.

These gages are designed for operation on d-c voltage. The length of the connecting leads between the indicator and transmitter has no effect on the readings. By referring to figure 5 you can refresh your memory on the wiring of a d-c selsyn system. Remember, it can be a two- or a three-wire system.

Indicators of d-c selsyn fuel gages are made with from one to four indicating elements—each indicating the contents of a separate fuel tank. A typical indicator element has three main parts—a rotor, element coils, and a core. The core is made of laminated metal and is circular. The element coils, which are connected to the transmitter resistance, are mounted on this core and connected to each other in series. The rotor is a permanent magnet, attached to a shaft. A transmitter diagram is shown in figure 51.

The shaft is turned by the effects of the coil fields on the rotor magnet. The direction of the field depends on the currents sent out by the transmitter, as you have seen. Surrounding the magnetized rotor is a damping ring of copper, which is necessary to prevent the rotor from "hunting" or wavering before reaching its proper position.

A small bar magnet is attached to the bottom of the frame of each element. When the element is not connected to a source of power, this magnet pulls the rotor to a position that holds the indicator pointer off the dial scale. Thus, when disconnected, an element indicator will have an off-scale reading.

Some indicators having two elements are made with these

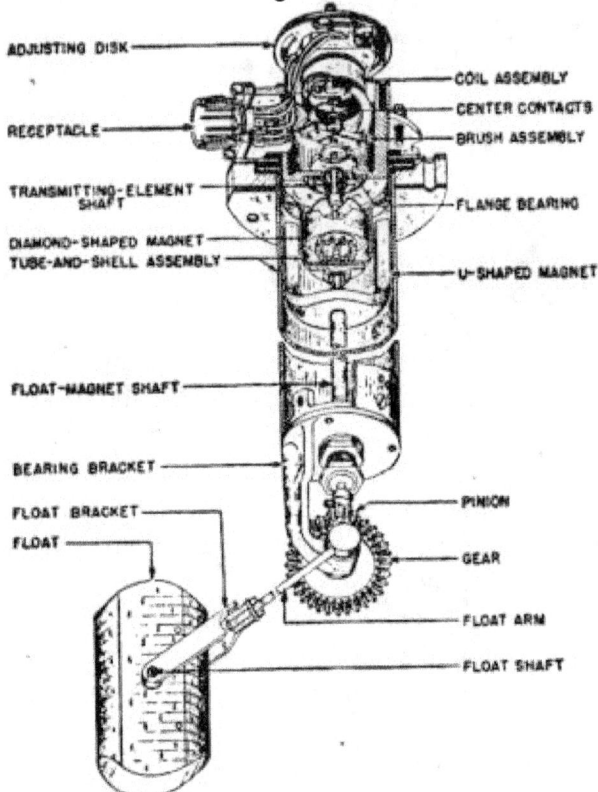

Figure 51.—Diagram of d-c Mltyn transmitter.

elements back to back and shielded from each other. The shaft of the rotor assembly of the front element in these models is made hollow so the shaft of the back element can extend through it. Some have two pointers that are pivoted at the center of the dial, but turn in opposite directions and in front of separate calibrated scale markings. Others have two pointers—one shorter than the other turning in the same direction over concentric dial scales which indicate the fuel levels in different tanks. Indicators with three or four elements have separately mounted pointers, each of which has its own gage scale on the dial.

The indicator elements used in any one model are identical except for the location of the small magnet which holds the pointer off the scale when the power is off, and the color of the internal lead wires. These colored leads are used for easy identification in making connections to the plug-in receptacles at the back of the instrument.

A transmitter is mounted in the top or side of the gasoline tank, and is inserted into the tank through an opening made for the purpose. If you get the job of installing a transmitter, be sure that the tank is empty and that the float or float arm does not strike any baffle inside the tank. The float has to be free to move throughout its full arc or else the gage cannot register through its complete range.

A transmitter unit is secured in position on its tank by means of screws which pass through the flange of the transmitter, or by means of a threaded flange adapter and tank fitting, depending on the model. You must be careful that the transmitter is installed in such a position that the float arm turns the pinion gear in a counterclockwise direction when the fuel level rises.

You will have to make an adjustment of the transmitter for empty-tank condition before installation is complete. On one of the standard models, for example, with the float resting on the bottom of the empty tank, the transmitter must be connected with its indicator (by means of the leads) according to the wiring diagram.

The four screws on the top of the transmitting element of this type of transmitter must be loosened and the adjusting disk rotated until the pointer reads zero. Then you re-tighten the screws. If you find it impossible to get an empty tank adjustment this way, rotate the adjusting disk so that it is at its midpoint of travel, and tighten the four screws slightly to hold it in place. Note the indicator reading at this point. Then remove the six screws which hold the transmitting element to the transmitter mounting flange. Remove the element and rotate the shaft to a position where the indicator reading is the same as the first reading you noted.

With the shaft in this position, make a small pencil mark on one end of the diamond-shaped magnet and a mark on the case of the element directly opposite this end of the magnet.

Then grasp the transmitting element. Push the magnet against the spring and rotate the shaft until the cross-pin alines with the notch in the magnet which makes the indicator read nearest to zero when the marks on the magnet and on the case are realined. This makes a rough empty-tank adjustment. You then proceed as outlined before to make the adjustment exact. Procedure and adjustments will differ with the various models of transmitters.

LIQUIDOMETER GAGES

A type of fuel gage very similar to the d-c selsyn in construction is the electrically operated Liquidometer. Tank units in the latter are available in two forms—a float-andiron type, and a direct-lift float type. In both types, the movement of the float is carried to the electrical part of the transmitter by direct mechanical linkage. No gears are involved.' The linkage works through a rocking metal bellows which at the same time acts as a seal between the electrical parts and the fuel tank contents.

The tank units are constructed so that the movement of the pointer over the face of the calibrated indicator dial can easily be adjusted to indicate "empty" when the float touches the tank bottom, and "full" when it touches the top— without removing the tank unit.

The fuel tank gaging requirements of various airplanes are quite different, as you know. This accounts for the fact that indicators are available with a number of different pointer and dial arrangements. Some electrical Liquidometer indicators, for example, are furnished with a single pointer which moves over the face of a 90°, 120°, or 300° calibrated scale. Others have a number of scales with a pointer for each, mounted in a single indicator housing. Each of the electric mechanisms in an indicator housing is shielded, so they can operate without electrical interference.

One model of a single-pointer indicator has a mechanism made up of a moving-coil type of ratio resistance bridge, arranged to move in the field of a permanent magnet. There are two coils in the moving part, and these coils carry the

274033°

pointer. The variable resistance in the tank unit controls the ratio of the current flowing through the two coils. This current ratio is what determines the position of the coils with respect to the poles of the permanent magnet. The fuel level, of course, determines the resistance in the circuit by changing the position of the contacts on the resistance strip in the tank unit.

Another type of single-pointer indicator has stationary coils and a moving magnetic rotor. In such instruments, the rotor carries the pointer. Otherwise the principle of operation is the same, with the variable resistance in the tank unit controlling the current flowing through the stationary coils. These coils in turn produce a magnetic field which controls ' the turning of the rotor. Dual-pointer indicators and cross-pointer indicators use two of the moving rotor type of mechanisms.

When an airplane has a number of tanks of different capacities and a separate dial is needed for each tank, a dial-change indicator is frequently used. This type of instrument is a great space-saver, having only one mechanism in the indicator and using a single pointer. The important extra feature in this instrument is its selector switch equipment, which permits the operator to check the fuel level in any tank connected to the instrument.

How does a dial-change indicator work ? When the selector switch knob is turned to a given position so as to connect the tank unit of a certain tank to the indicator mechanism, a dial calibrated for that tank automatically comes into view behind the pointer. The instrument then operates in exactly the same Way as a moving-coil type single-pointer indicator. Shifting the selector switch knob to another position does three things—it disconnects the indicator from the first tank, connects it to another, and brings the proper dial into view for obtaining a reading on the newly connected tank.

Tank units, as has been pointed out, are made in two styles—the float-and-arm type, and the direct-lift float type. The former comes in several models (many of which are not discussed here) for use with the various types of indicators. Direct-lift units consist essentially of a long tube with a circular housing at one end and a locating socket at the other, as you see in figure 52. Inside the tube is a large float. Attached to the float is a fixed roller running in a spiral guide slot in the tube. As the float rises or falls, it transmits a rotating motion to a central shaft. The turning of this shaft changes the position of a contact arm on the resistance strip in the housing, varying the current ratio delivered to the indicator coil in the same way as other types of tank units operate.

If you are installing a direct-lift tank unit, make sure that the edge of the mounting hole in the tank is free from burrs or metal protrusions. Mounting screw holes in the tank flange should be "blind"—that is, they should extend only part way through the flange. Gasket paste should be applied, and a gasket placed on the tank flange before installation of the tank unit itself. Be certain that the lower end of the direct-lift unit enters the locating socket at the bottom of the tank. Then turn the unit until the conduit fitting is in correct position, and screw the unit in place. Finally, attach the conduit and leads according to the proper wiring diagram.

When you install a float-and-arm unit, notice that the float arm is fitted with an eye through which a loop of string can be passed, as in figure 53. After the unit has been put in place, the float should be moved up and down by means of the string to make sure it strikes the top and bottom of the tank and has uninterrupted movement. As soon as you are satisfied that the float has proper clearance, you should secure the unit in place with the mounting screws. But don't remove the loop of string before the stroke-setting operation.

After all units which make up the complete gage are installed and connected according to the correct wiring diagram, you are ready for stroke setting. Frequently the indicator is located in a place where you can't see it while you're adjusting a tank unit, so be certain to have someone else along to help you if such is the case. Your helper should

Figure 52.—Direct lift tank unit.

Figure 53.—Float-and-arm tank unit.

be stationed at the indicator unit, and have his signals straight so you'll both know what they mean.

Now take a look at figure 54 which is a diagram of the potentiometer in the housing-of a tank unit. Before going to work on setting the stroke, be sure that the adjustment screws C and D are turned so their slots line up together as illustrated. The contact arm should be centralized so that it travels an equal distance from each end of the resistance strip. This can be done by hand, since the contact arm is held to its pivot simply by a friction fit.

When the switch is "ofty' the pointer will be below the "empty" mark, resting against its stop pin. When you turn the switch "on" (or to the position for the tank you're cheek-

Figure 54.—Tonk unit potentiometer.

ing) the pointer should move. If it doesn't stand at the "empty" mark, turn adjustment screw C until it does. Then move the float to touch the top of the tank by means of that loop of string, and adjust the screw labeled D until the pointer indicates the "full" position. Try these top and bottom positions two or three times, and readjust the screws if necessary.

AUTOSYN FUEL FLOWMETERS

When flying it's important to know how fast an airplane's engines are consuming the fuel supply. An instrument providing this information is the autosyn fuel flowmeter. This measurement is transmitted electrically to the panel-mounted indicator. The use of electrical transmission eliminates the need for a direct fuel-filled line from engine to the instrument panel. Possibilities of fire hazard and mechanical failures are greatly minimized. The fuel flow transmitter does not measure the quantity of fuel in the tank. As you. would probably guess, the fuel flowmeter is quite similar to other autosyn instruments already considered. It consists of a fuel flow transmitter and a standard autosyn indicator.

The transmitter is a sort of two-in-one unit, being made up of a fuel-measuring mechanism (or meter) and an autosyn motor. These parts can be separated from one another for maintenance purposes, but are joined together as a single •assembly for installation.

How does the fuel-measuring mechanism work? Fuel on its way to the carburetor enters a spirally cut chamber in the measuring unit and strikes a pivoted vane. The vane shaft extends up through the fuel chamber cover and supports a bar magnet. One end of a spiral spring of measured strength is attached to the shaft, and tends to oppose the turning induced by the fuel flowing against the vane.

The autosyn motor assembly has a ring magnet assembly attached to its rotor shaft. Thus the ring magnet and the bar magnet form a magnetic coupling. Their polar attraction "locks" them together so that when the bar magnet is turned it carries the ring magnet around with it. This, of course, turns the transmitter rotor to a given position—depending on the push delivered on the fuel chamber vane by the flowing fuel. The transmitter thereupon operates like any other autosyn transmitter. The impulses it sends to the indicator motor positions the indicator rotor, and a reading in pounds per hour is obtained on the dial.

The fuel flow transmitter is equipped with a relief valve which automatically opens and bypasses the instrument whenever the fuel pressure drop across the instrument gets too high—as it may during takeoff, for example. At such time, only part of the fuel flows through the metering portion. As soon as the pressure across the instrument falls below the value at which the relief valve is set, the valve closes and the flowmeter again goes to work.

An autosyn fuel flowmeter can be used with either a pressure- or gravity-feed fuel system. The transmitter unit must be located in the fuel line—between the fuel pump and the careburetor—when used with a pressure system, so that only the fuel flowing into the carburetor passes through the metering unit.

When you install a flowmeter transmitter, make all fuel connections with flexible tubing. Connect the line from the strainer to the transmitter vent marked "in" and the line from the vent marked "out." Pipe and tubing connections must be coated with antiseize, and you'll have to see that the lines aren't clogged with excess compound.

CAPACITOR TYPE FUEL QUANTITY GAGE

v

The capacitor type fuel quantity gage is an electronic system for measuring quantity of fuel in aircraft in pounds. The operation of the older models of the Simmonds systems requires a 28 volt d-c power supply. Newer systems are powered by alternating current. Each system consists principally of a tank unit for each tank, power units, and an indicator. The components of a capacitor type fuel quantity gage system are shown in figure 55, and discussed in the following paragraphs. This system is the Simmonds system, but much of the discussion is applicable to other systems as well. Consult AN publications for detailed descriptions of the other systems.

In a Simmonds system the tank unit is designed to be mounted inside a fuel tank with a coaxial cable connecting it to the bridge circuit of the capacitor system. The tank

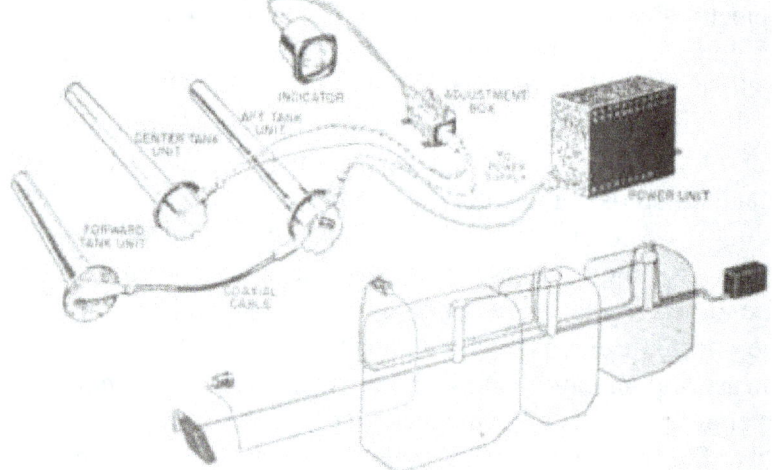

Figure 55.—Capacitor type fuel quantity gage system.

unit consists of two aluminum alloy tubes, the smaller assembled concentrically within the larger tube. Space between the tubes is approximately y 32 of an inch, except at the bottom where the inner tube is tapered to drain off water droplets which might be formed by condensation. A drilled hole in the top cap vents the air, while a hole in the bottom cap permits fuel to enter the unit. The two tubes function as a condenser and work on the principle that the dielectric of the insulating medium between condenser plates in an electric circuit varies the electric capacity of the condenser. This combination of two metallic plates in the cell unit is so situated that when the cell is full, only fuel is between the tubes. As fuel is used, an increasing proportion of air replaces it, until there is only air between the tubes. The proportion of fuel or air between the tubes of the cell unit affects the electrical capacity of the unit (because the dielectric constant of fuel and air differs) so that the capacity may be said to be a function of the fuel level at the particular location of the cell unit. The changes in the electrical capacity of the cell unit are measured by a capacitance measuring unit, or power unit, in which an electronic circuit measures the value of the electric capacity of the cell unit in terms of small direct current. These small direct currents are in turn fed to a ratiometer type indicator.

The power unit supplies power for the indicating circuits by translating condenser capacity changes made in th'e tank unit and registering them on the ratiometer-type indicator. The fuel quantity indicators are ratiometer-type indicators. The indicator is essentially a moving-coil, stationary-magnet type instrument, with a moving element consisting of two coils moving in the air gap of a permanent magnet. The large coil is a deflecting coil which functions as the moving coil of a galvanometer, and the small coil is a restoring coil which is so connected as to oppose the action of the large coil. Three hairsprings are used on the moving element. They act principally to provide sufficient torque to return the pointer to its mechanical zero when the ratiometer is disconnected. This mechanical zero position is off scale at the low end of

the arc. The moving coils and the various resistors and calibrating potentiometers of the indicator are connected in the form of a modified bridge circuit arranged so that when the voltage ratio is one to one, the pointer will indicate zero.

For accurate indication, full and empty adjustment of the fuel gage system must be accomplished on each airplane at initial filling of new fuel cells and at any time thereafter that a component of the system is replaced. Upon completion of each adjustment, entry must be made on the adjustment card. This data permits routine tests to be made without repetitious measurement and regardless of the quantity of fuel present in the cells.

To adjust the fuel quantity gage system for empty adjustment, proceed as follows:

1. Make certain that fuel cells served by the channel to be adjusted are empty of fuel available to engines.

2. Apply power and allow approximately 5 minutes for power supply tubes to reach operating temperature.

3. Set indicator to zero with the respective empty adjustment screw of the adjustment box.

4. Connect the capacitance decade at the respective rectifier head on the tank unit.

5. Apply capacitance until indicator is again set to zero. Do not change adjustment box setting. Arithmetically add the capacity of the adapter to the decade reading and enter this figure on the adjustment card.

To adjust the fuel quantity gage system for full adjustment, proceed as follows:

1. Fuel density is a variable greatly affected by temperature and specific gravity. The weight of fuel required to completely fill a fuel cell may differ from one filling to another. With correct scale alinement, the weight of fuel indicated is the actual weight of fuel in the cell.

2. Fill cells served by channel to be adjusted, measuring volume in gallons on an accurately calibrated fuel meter.

3. Determine specific gravity of fuel with a hydrometer.

4. Calculate the weight of fuel serviced by multiplying the fuel contents in gallons by the specific gravity and then by 8.32. Enter this weight on adjustment card.

5. Apply power and, by means of respective frill adjustment screw of the adjustment box, set indicator to fuel quantity determined to be in cells.

6. Connect capacitance decade and apply capacitance until indicator is again set to fuel quantity determined to be in fuel cells. Do not change adjustment box setting. Arithmetically add the adapter capacity to the decade reading, and enter this figure on the adjustment card.

If the indicator reads below scale with no current at the indicator, the power unit is not supplying a-c current or the lead from the power unit is faulty. The lead or power unit, whichever is faulty, must be replaced. If the indicator reads low with full fuel cells, there is an open in the tank unit, cable, or one of the tank units is not grounded. The unit or cable must be replaced and the tank unit grounded, depending upon the cause. If the indicator reads zero at all times, there is probably a short in the circuit. Circuits and leads must be checked and replaced if found to be faulty.

QUIZ

1. A typical d-c selsyn indicator element has three main parts. These are , , and

2. The shaft of the d-c selsyn indicator is turned by the effects of the on the

3. A small is attached to the bottom of the frame of each selsyn element.

4. Be sure that the tank is when installing a transmitter.

5. If you are installing a direct-lift tank unit, make sure that the edge of the mounting hole in the tank is

6. When you install a float-and-arm unit, the float should be moved up and down to make sure it .

7. To indicate how fast an airplane consumes its fuel, a(n) — is installed.

8. Does the fuel flow transmitter indicate the amount of fuel in the tank?

9. The fuel flow transmitter is equipped with a. which automatically opens and bypasses the Instrument whenever the fuel pressure drop across the instrument gets too high.

10. An autosyn fuel flowmeter can be used with either a or a fuel system.

11. A pulls the indicator rotor to a position that holds the indicator pointer off the dial scale when the power is disconnected.

12. The float arm of the liquidometer should turn the pinion gear In a direction when the fuel level rises.

13. The electrically-operated liquidometer is very similar to the

14. Tank units are made in two styles: type, and the type.

15. When you install a flow-meter transmitter, make all hose connections with tubing.
a. Copper.
b. Flexible.
c. Aluminum.
d. Self-sealing.

16. Most capacitor type fuel quantity gages require a power supply.

17. The fuel quantity indicators of the capacitor type fuel quantity gage are type indicators.

18. The length of the connecting leads between the indicator and transmitter of a d-c selsyn fuel gage

19. A prevents the indicator rotor of a selsyn fuel gage from "hunting" or wavering before reaching its proper position.

20. When disconnected, an element indicator will have
a. A maximum scale reading.
b. An off-scale reading.

c. A middle scale reading.

d. A zero scale reading.

21. The float of a fuel gage has to be free to move through its full arc so that the gage

22. Each liquid meter gage has shielded electric mechanisms in the indicator housings so that they can .

23. When an airplane has a number of tanks of different capacities and a separate dial is needed for each tank, is frequently used.

24. The capacitor type fuel quantity gage is a(n) system for measuring fuel in aircraft in .

25. The two tubes of capacitor type fuel quantity gage act as a .

CHAPTER

COMPASSES

From the beginnings of recorded history, man, especially the seafaring man, has been vitally interested in where he was going. Countless devices and methods have been invented and devised to accomplish this.

In the present era, with its supersonic speeds, accurate determination of direction has become increasingly important. An error of but a few degrees in a space of minutes will rocket the modern mariners of the air many miles off their course.

DIRECT-READING COMPASSES

During the infancy and childhood of aviation, direction of flight was determined within the aircraft chiefly by direct-reading compasses. Even today, in what might be called the adult stage of aviation, the direct-reading magnetic compass finds its use as standby compass.

Direct-reading magnetic compasses used in Navy airplanes are of two general types. One type is mounted on the instrument panel for the use of the pilot, and can be read like the dial of a gage. The other type is a top reading navigator's compass, and it is mounted face upward, usually on the navigator's table or some other horizontal surface. Both types can be seen in figure 56.

The fundamental parts of both types differ somewhat in appearance but both operate in almost identical ways. A nonmagnetic metal bowl, filled with liquid, contains a card which provides the means of reading compass indications. A set of small magnetized bars—or needles—is fastened to this card. The card-magnet assembly is suspended on a

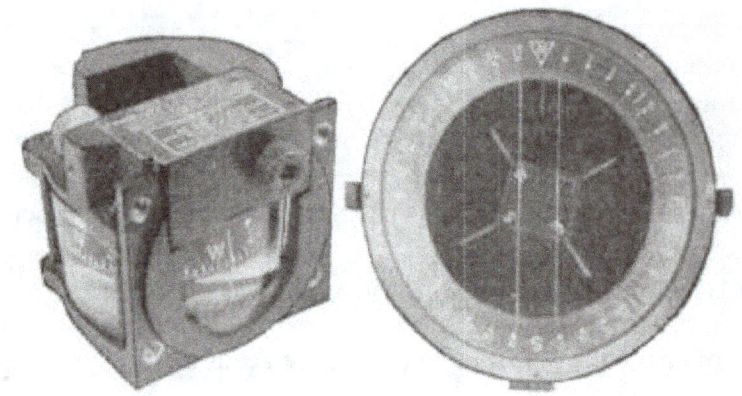

Figure 56.—Left, pilot's compass; right, top-reading compass.

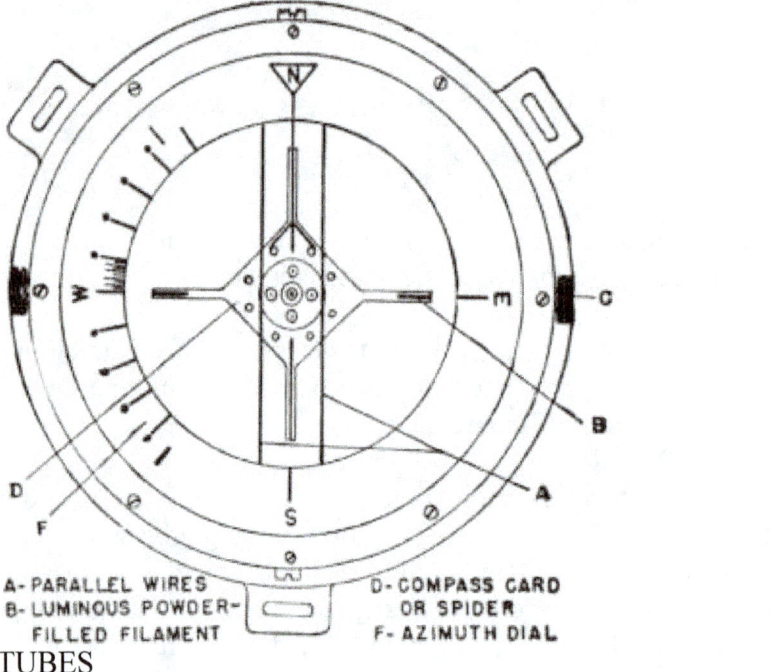

A- PARALLEL WIRES
B- LUMINOUS POWDER-
   FILLED FILAMENT
TUBES
C- A2IMUTH RING LOCK SCREW

D- COMPASS CARD
   OR SPIDER
F- AZIMUTH DIAL

Figure 57.—Sectional diagram of top-reading navigator's compass.

jeweled pivot which allows the magnets to aline themselves freely with the North-South direction of the earth's magnetic field. The compass card, and a fixed-position reference marker called the lubber's line, are visible through a glass window on the side or top of the bowl.

An expansion chamber is built into the compass to provide for expansion and contraction of the liquid resulting from altitude and temperature changes. The purpose of the liquid is to dampen— or slow down—the oscillations of the card caused by vibration and changes in the airplane heading. If suspended in air, the card would keep swinging back and forth and be difficult to read. The liquid also buoys up the float, reducing the weight and friction on the pivot bearing. General details of the top-reading compass are shown in figure 57.

READING THE COMPASS

Instrument-panel compasses for Navy airplanes are available with cards marked either in steps of two or five degrees. Top-reading compasses usually have graduations marked in two-degree steps. You don't have to make any settings on an instrument-panel compass to find out the approximate heading of the airplane. Such a compass indicates continuously, and the heading may be read by looking at the card in reference to the lubber's line through the bowl window, as shown in figure 56.

The top-reading compass also indicates the magnetic heading of the airplane when its grid lines are set in alinement with the card pointer. This type of compass is fitted with a verge ring, which may be rotated to bring the grid lines into the desired position. The heading is then read directly from the graduated scale of degrees on the verge ring with reference to the lubber's line, as in figure 56.

INSTALLATION

You can't install an airplane compass any place and expect it to function properly. It simply won't behave unless its surroundings suit its temperamental nature. Special locations for installing compasses are provided by airplane

manufacturers, and these locations weren't chosen by guesswork. So don't install compasses elsewhere unless relocation is necessary and has been specifically authorized.

The pilot's instrument-panel compass and the navigator's top-reading compass must both be mounted so that a line passing through the card pivot and lubber's line is parallel to the longitudinal (fore-and-aft) axis of the airplane, and the card-pivot supporting post must be perpendicular to the horizontal when the airplane is in level flying position.

As you'd expect, the screws and brackets used for mounting compasses must be of nonmagnetic metal. Brackets are usually made of aluminum or brass. Remember when you're mounting a top-reading compass that the compensator adjusting screws must remain easily accessible. The "why" of this will be apparent later.

You have to be especially careful when youre installing any magnetic compass to see that it's placed where disturbing magnetic fields in the vicinity will affect it the least. Disturbing magnetic fields can be set up by nearby electrical equipment, radio gear, electrically operated armament, metal structural members of magnetic material, electric wiring, and many other items of otherwise innocent appearance. A compass can be compensated for a reasonable amount of permanent magnetism near it, but no method of compensation will eliminate the effects of variable magnetic fields. The maximum allowable deviation of a compass before it's compensated must not be more than 12°, or more than 5° after compensation.

It's necessary to remove a compass and replace it with a serviceable instrument under the following conditions—

1. When card markings are illegible because of fading, discoloration, or loss of luminous paint.

2. If card doesn't rotate freely when the airplane is in normal flying position. This can be checked by deflecting card with a small permanent magnet.

3. If bowl is cracked, or if mounting frame or lugs are broken.

4. When compass is erratic or doesn't respond after proper efforts to compensate it (unless erratic behavior is caused by compass location only).

5. If lubber's line is loose or misalined.

6. If compass needs more liquid, or requires bench tests and disassembly for major repairs or adjustment.

COMPENSATING THE COMPASS

Aircraft magnetic compasses are equipped with devices called compensators, which provide a means of correcting for deviation errors. As was pointed out before, you can't eliminate all such errors, but you can cut them down to a minimum by the process called swinging, which is part of your job.

Compensators are of two types. One is known as the universal screw-type, and consists of an assembly having a group of small compensating magnets permanently installed in it. Adjustments to change the compensating effect of the assembly are made by means of two adjusting screws—one for north-south compensation, the other for east-west. The other type of compensator employs small, loose magnets which are placed in special chambers on the compass as needed. One such chamber is placed so that its magnets make east-west corrections; the other (at right angles to the east-west chamber) corrects north-south deviation.

Before you start the swinging procedure, make certain that all magnetic equipment is secured in the position it will occupy in normal flight. Also be sure that neither you nor anybody else who is to read or be near the airplane compasses during swinging operations has any magnetic materials on his person. Magnetic materials include tools, pocket knives, mechanical pencils, wrist-watch or dog-tag bracelets, eyeglasses, officers' caps, badges, and many other things—so be careful to think of all the possibilities and eliminate them. Remember, too, that a nonmagnetic screw driver must be used in adjusting universal compensators.

The actual swinging of a compass may be done in one of several ways, but in your work you'll be chiefly interested in

274033*—64 9

ground swings. Ground swinging is usually done with the airplane at rest on a compass rose —as shown in figure 58. A compass rose is much like an oversized card from a navigation compass, and is usually laid out or painted on the ground at most air bases. The directions shown by it are magnetic directions—that is, the north arrow points toward the earth's north magnetic pole, but many compass roses also have a line showing true north.

Jacks, lifts, hoists, or any dolly needed on the ground swinging job should preferably be made of nonmagnetic material. Since that's often impossible if they have magnetic metal in their parts, they must be tested for their effects on the compasses by moving them about the airplane in a circle at the normal distance that would be between them and the instruments. Don't use equipment that causes a change in compass readings of more than one-quarter degree.

Trucks, automobiles, railroad cars, and other airplanes contain magnetic metals and should not be within the swinging area where they'll have magnetic effects on the compasses of the airplane on which you're working.

There's no point in wasting your time—and that of other people—swinging a compass that's not in good condition. So, before you start, examine the compass to make sure that the liquid is clear and at the proper level, and that the card assembly is level and turns freely when the airplane's tail is lifted to flying position.

Now, let's consider the swing itself. Suppose that you're stationed aboard the airplane as a compass observer during the operation. First, set the compensator so that it has no effect on the main compass magnets. With a loose-magnet compensator this is done by removing all loose magnets from their chambers. Universal screw-type compensators are set for zero effect by turning both adjusting screws until the dots on the screws are matched up with the dots on the compensator case.

The airplane is then placed directly on a south magnetic heading over the compass rose—with the tail raised to level

flying position. Next, note tiik compass reading and make a record of it. From this reading, it's simply a matter of algebraic subtraction (or subtraction of numbers having plus and minus signs) to determine the deviation on the south heading. The deviation is the algebraic difference between the magnetic heading and the compass reading.

Figure 58.—Compass rose, with airplane on south heading.

Following this, the airplane is placed on a west heading. Again note the compass reading, and figure out the deviation, or difference between the magnetic heading and what your compass tells you.

The next heading to which the airplane should be turned is magnetic north. After you've taken the compass reading on this heading, and found the deviation, you have a little problem to work out—easy but important.

Here's an example. Suppose that on the south (180°) heading your compass reads l75i/ 2 °- In other words, the compass reads 4y 2 0 less than it would if there were no deviation. You would record this as a deviation of +4A/ 2 ° (180°-175%°). Kemember, if the compass reading is too low, the deviation is plus; if the reading is too high, the deviation is minus.

Now, suppose that on the north (000°) heading, your

compass reads 006°. Such a reading is 6y 2 ° too high, so you would record this as a deviation of ~8%° (000°-006i/ 2 °).

Your next job is to work on the coefficient of north-south deviation by subtracting, algebraically, the deviation on south from the deviation on north, and dividing the remainder by 2—like this:

(-6y 2 °)-(i4y 2 °)_-ii o KUO

Your airplane, you recall, is still on the north heading, and the compass reads 006°. But you have now found the coefficient of north-south deviation to be -5°, so you adjust the north-south compensator by this much, and your reading on the north heading will now be 001°. This adjustment also corrects the south deviation by the same amount, so that on a south heading your compass will now read 181°. The coefficient of north-south deviation, which turned out to be —5y 2 ° in this case, is called coefficient a C" in Uncle Sam's military circles. As you can see, it is the average of the deviation on the two headings. Both the Air Force and Navy have agreed on this and certain other convenient short names, for uniformity in compass-swinging practice.

If the compensator is the loose magnet type, the adjustment for north-south deviation is made by inserting the necessary number of magnets into the lateral (athwartship) chamber of the compensator. If, however, the compass has a universal compensator, you make the adjustment by turning the north-south (N-S) compensator screw.

You still have to take care of east-west deviation. The airplane must be turned so that its heading is magnetic east, according to the compass rose, and you make a record of the compass reading on that heading. And now you're ready to figure out the coefficient of east-west deviation, otherwise known as coefficient "B."

Assume, for example, that the compass reads 276° when the airplane was on the west (270°) heading, and reads

exactly 90° on the east (90°) heading. Coefficient "B" is found by algebraically subtracting the deviation on west (—6°) from the deviation on east (0°) and dividing by 2, like this:

While your airplane is on the east heading, you adjust the east-west (E-W) compensator to add 3° to the compass reading. This reading becomes 93° on the east heading and, of course, the compass would read 273° on a west heading. The adjustment is made by turning the E-W screw on a universal compensator, or by adding the necessary magnets in the longitudinal (fore-and-aft) chamber if the compass compensator is of the loose magnet type.

Leaving the airplane on an east magnetic heading, you next compute an overall deviation correction, based on what is called coefficient "A." This coefficient is equal to the algebraic sum of the compass deviations on all four cardinal headings (north, east, south, and west), divided by 4, like this:

Instrument-panel compasses must be compensated for coefficient "A" if it amounts to 2° or more in either direction, and horizontally mounted compasses if it's more than 1°. Whichever type you're working with, then, would have to be further corrected in this particular case, since the overall deviation is —2°. But this time leave the magnetic compensation devices strictly alone. Compensation for coefficient "A" is done simply by moving the instrument in its MOUNTING.

Compensation of panel-mounted compasses for coefficient "A" can be accomplished either by a slight realinement of the whole instrument panel, or by turning the compass a little with relation to the front of the panel and placing washers or spacers under its mounting screws. Horizontal

0°-(-6°)_+f>° 2 ~ 2
+ 3°.
(-6y 2 o)+0 o +4i/ 2 o +(-6 o) _ (-8°) 4 4
2°.

compasses can be compensated by loosening the mounting screws and rotating the instrument the required amount before retightening the screws. And that does it!

After compensation is completed, the airplane must be swung again on eight equally spaced headings such as, for example, 0°, 45°, 90°, 135% 180°, 225°, 270°, and 315°, and the compass readings recorded for each heading on a compass correction* card. This correction card is then mounted on the instrument panel or navigator's table as close to the instrument as possible. It thus is always available for ready reference, and will tell the pilot or navigator the comparative compass headings and magnetic headings for these eight readings.

REMOTE INDICATING COMPASSES

Keeping magnetic compasses from being seriously affected by magnetic metal and other disturbing influences is a problem that has kept airplane designers gnawing their nails for years. The fighting airplanes of today, heavily fitted with steel armor and guns, make the situation even worse, because armor and armament are heaviest around the cockpit, and that's where the compass must be read.

Airplane men have licked the problem by using remote-indicating magnetic compasses in armed and armored aircraft. The remote-indicating magnetic compasses found in many Navy fighting planes are of the magnesyn type. Such a compass has its magnetic "needle" in a transmitter which can be installed at any desirable point—as far away from disturbing effects as the size of the airplane will allow.

The flux-gate pickup unit—or "pickup"—is the part of the transmitter that passes along the "message" from the directional magnet to the indicator. Look again at figure 60, where the pickup (at left) is shown as a winding on a ring-shaped core of metal, and the directional magnet as a round disk marked N and S. At points A and B, alternating current from the a-c power supply is introduced into the coil, setting up an alternating magnetic field. The directional magnet, of course, being mounted inside the pivoted float, alines its poles with the earth's magnetic field.

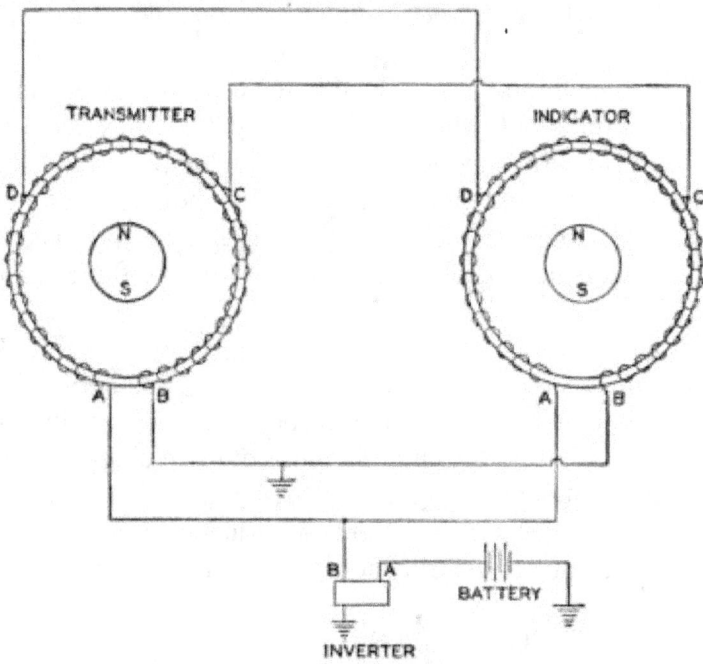

Figure 59.—Diagram of Magnesyn compass connections.

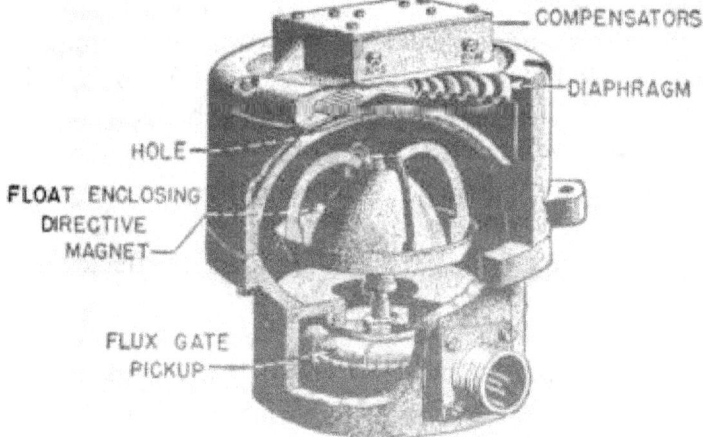

Figure 60.—Cutaway view of Magnesyn transmitter.

When the applied alternating current reaches its highest value in either direction during one alternation, it effectively shuts out the magnetic field of the permanent directional magnet rotor. When the current drops to zero at the end of each alternation, however, the field of the permanent magnet comes into its own again, setting up what amounts to another frequency in the pickup coil. The amplitude of this frequency is different for each direction in which the airplane may head.

The indicator is similar in principle to the transmitter, except that it is shielded from the earth's field. Its permanent magnet, therefore, is not affected by the earth's field, but is affected by the field of the indicator ring winding. This indicator winding receives alternating current from the same source as the transmitter coil.

Generally speaking, remote-indicating compasses operate on alternating current—much like autosyn instruments. The Magnesyn compass consists of a transmitter and one or more indicators, connected electrically as diagrammed in figure 60. The principal parts of a transmitter are the compensator (for making deviation corrections, the same as with a direct-reading compass), the expansion chamber, the bowl, the magnet-enclosing float, and the fluxgate pickup unit. You can identify these various parts in the cutaway illustration.

The expansion chamber of the transmitter is connected to the bowl by several small holes, and allows for expansion of the liquid with which the bowl is filled. It is covered by an expansible diaphragm. The bowl itself is nearly spherical, and the liquid in it is called damping fluid. The float,, which contains the directional magnet (or "needle"), is mounted on a pivot with a jewel bearing and is surrounded by the damping fluid. The upper part of the float has four damping vanes connected to it. The float is not connected mechanically to the fluxgate pickup unit, but affects it only

MAGNETICALLY.

The coils are connected by wire taps (D-D and C-C), so that the second harmonic voltage—set up by the directional. magnet—is identical at corresponding positions in both coils.

The strength of this voltage varies around the circumference of the pickup ring in the transmitter—the strength at any given point depending on the direction in which the transmitter's permanent magnet lies. The magnet in the indicator, free from the influence of the earth's field, assumes the same position (relative to its coil) as the directional transmitter magnet takes with respect to its coil.

GYRO STABILIZED FLUX GATE COMPASSES

You have read about the magnesyn remote-reading compass in the previous section. In the magnesyn compass, the magnetic needle and the transmitter can be placed at a considerable distance from the cockpit in order to avoid disturbing factors. Otherwise, such a compass has the same weaknesses and shortcomings as a direct-reading compass, because it is not stabilized.

The gyro stabilized flux gate compass not only is remote reading, but gets rid of most of the other annoying habits of ordinary magnetic compasses. How ? By stabilizing the transmitter element through the use of a gyroscope and thus eliminating the major errors and oscillations that pop up in magnetic compasses as a direct result of tipping during turns or rough air flying. Correct magnetic readings are thus continuously obtained on all headings, unaffected by turns, banks, climbs, dives, yawing about a given course, and bumpy weather.

The gyro flux gate compass consists of a transmitting unit, a master indicator, an amplifier, a remote caging unit (either. manual or electrical), secondary indicators or repeaters and, when required, an inverter. Inside the transmitter case is an electrically driven, vertical-seeking gyroscope, beneath which is mounted the flux gate element. This element transmits a signal induced in it by the earth's magnetic field, hence the gyro stabilized flux gate compass is known as an earth inductor compass. The signal transmitted by tho flux gate varies according to the direction in which the flux gate lies in the earth's field.

This signal from the transmitter is fed into an autosyn

in the master indicator. As a result, a continuously reading remote indicator gives the position of the flux gate element with reference to the horizontal component of the earth's field at

all times. The master indicator is constructed so that corrections for magnetic variation may be introduced. Another built-in feature is a mechanical correction device by which compensations can be introduced to correct for
DEVIATION.

Figure 61.—Master indicator, gyro flux-gate compass.

A pointer on the master indicator, as you see in figure 61, gives the corrected reading; a dial visible through a small cutout, gives uncorrected readings. You thus have both for comparison. Readings are transmitted to repeater indicators like those in figure G2 by a transmitting magnesyn, also mounted in the master indicator. The readings you get on the repeater indicators arc corrected for deviation, and, if you have introduced a correction for variation by turning the outer dial of the master indicator, these corrections will also show on the repeater indicators.

In order to provide sufficient power to turn the induction motor in the master indicator, an amplifier is included in the system. This is a vacuum tube affair, to step up the signal originating in the flux gate. The amplifier also provides current to put the flux gate in action and to run the induction motor in the master indicator.

Figure 62.—Repeater indicator.

A different type of apparatus is used in this compass to keep the transmitter gyro erect than is used with the horizon indicator gyro. This erection mechanism consists of a small steel ball that runs in a circular track on the top of a cap on the gyro frame. A slotted disk is mounted over the track and the steel ball is free to run in its track within the limits set for it by the slot.

A small circular magnet, attached to the rotor shaft of the gryo, turns inside a cup which is a part of the erection disk. The magnet is just strong enough that when the gyro is running level and at its correct speed (approximately 10,500 r.p.m.) the disk of the erection mechanism is dragged along at about 40 r.p.m. As the disk turns, of course, it also carries the steel ball with it. When (he axis of the gyro is vertical,

the ball runs evenly in its track and there is no unbalanced force acting on the gyro. When the gyro axis is tilted, the track of the rolling ball also is tilted. The speed of the ball is increased on the downhill side and decreased on the uphill side of the track.

When the gyro heels away over, the weight of the ball is sufficient to prevent the erection mechanism from turning at first. The ball will be carried part way up the uphill side of its track by the force of the turning magnet and will therefore exert a force on the gyro approximately at right angles to the high side. Now, remembering the rule that governs the precession of gyroscopes, you will see that the force exerted by the ball tends to make the gyro right itself by precession. In principle, the rest of the transmitter mechanism of a gyro flux gate compass is very much like the magnesyn compass you have studied, although the construction is quite different.

The caging mechanism is an important part of the transmitter mechanism of a gyro flux gate compass. The gyro is caged by turning a caging knob, located at the bottom of the transmitter case. The knob is turned in a clockwise direction whether you-are caging or uncaging. Some installations have manually operated remote caging units. Others have caging units run by an electric motor and controlled by a switch.

The master indicator serves to combine the electrical impulses from the flux gate element so as to indicate the magnetic heading. It also serves as a secondary transmitter to send its readings to as many as five secondary indicators. The master indicator consists of a coupling autosyn, an induction motor, a transmitting magnesyn, an indicator pointer and dial, and mechanisms by which corrections may be made for variation and deviation. The coupling autosyn serves as the control unit of the master indicator and determines how far and in what direction the pointer should be rotated.

The small output of the rotor of the coupling autosyn is sent to the amplifier circuit where it is stepped up by means of a vacuum tube circuit. The resulting current is then used

TRANSMITTER

f

INVERTER

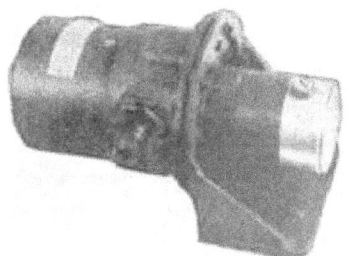

MANUAL CAGING CONTROL

MOTOR-DRIVEN CAGING CONTROL CAGING CONTROL SWITCH

Figure 63.—Gyro flux-gate compass parts and accessories.

to turn the induction motor. The induction motor supplies the mechanical power necessary to turn the moving parts of the master indicator. It is connected by gears to the pointer of the master indicator, and also to the rotor of the transmitting magnesyn which operates the repeater indicators in various parts of the airplane.

The amplifier unit is a jack-of-all-trades. It serves as a power distribution center to operate the gyro motor, the magnesyn system, and the vacuum tube circuits. It supplies the excitation current for the flux gate, amplifies the output current of the coupling autosyn of the master indicator, provides a supply of current for the master indicator induction motor, and serves as a junction box for the wiring of the entire compass system.

When the airplane in which a gyro flux gate compass is to be installed does not have its own supply of alternating current power, an inverter is introduced into the compass system to convert the current from the airplane battery-generator system into alternating current that the compass can use. The inverter consists of a 24 to 28 volt direct-current motor and alternator mounted on a common shaft generating 115 volts of 400-cycle current for operating the compass. An inverter is unnecessary when the airplane has its own supply of alternating current.

RADIO COMPASSES

Electronics has invaded all fields of endeavor in this age of electronics. Radio compasses guide aircraft to a transmitting station at its destination. While the radio compass is in use as a direction indicator, the pilot and navigator can also hear the station signals to obtain weather or other flight information. It has the additional use of a radio communication receiver while not serving as a compass.

The general functioning of a typical radio compass is outlined in figure 64. The directional loop receives a signal which passes through a stage of preamplification. The phase of the signal is retarded 90°, and then it is fed to both grids or a dual-triode tube used as a balanced modulator.

The modulated signal is combined with the signal from the sense, or nondirectional, antenna, and the result goes through a semiconventional superheterodyne receiver. The detected signal is then fed to two independent circuits, one

NO N 01 RECTI ON A L ANTENNA

Tj

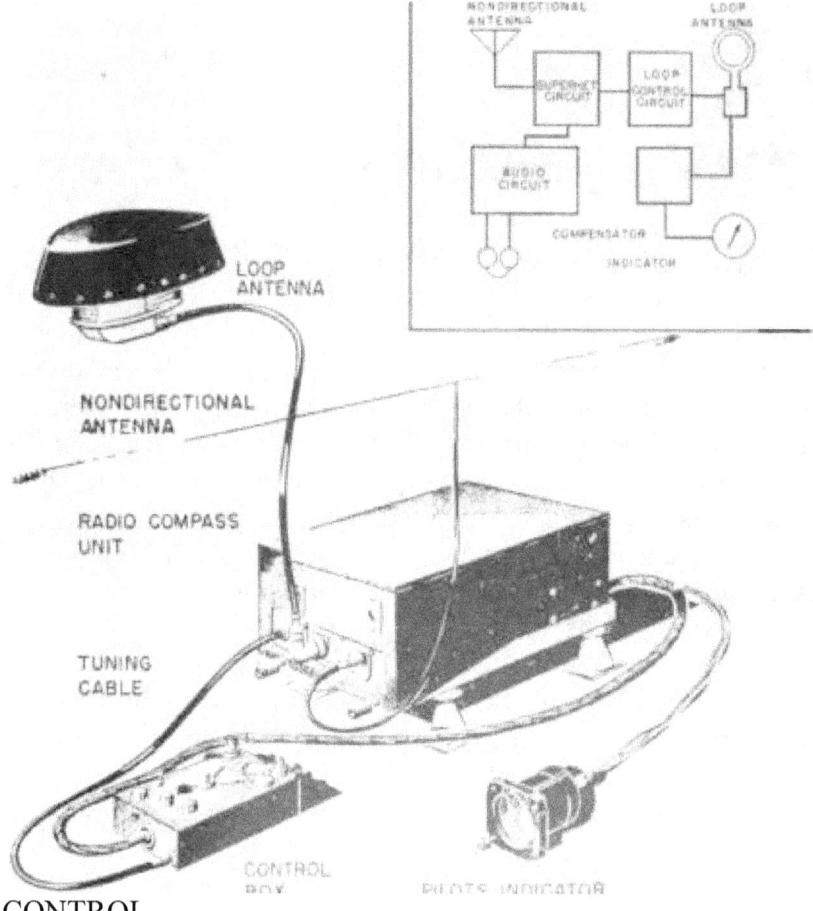

CONTROL
BOX PILOTS INDICATOR

Figure 64.—Functional block diagram.

for listening, and the other for automatic control of the loop drive motor.

The position of the loop is remotely shown on the indicators by an autosyn transmitter connected to the loop

through a "compensator." The compensator is a necessary addition to the system because the line of travel of the received radio wave is subject to distortion by the metallic structure of the airplane.

Since the loop is unable to allow for the sending station appearing in a false direction, the compensator is placed between the loop coil shaft and the autosyn transmitter, causing the transmitter to run behind or ahead of the loop.

The loop antenna is directional in that the voltage induced in the loop is maximum when the line of travel of the received radio wave lies in the plane of the loop coil as illustrated in figure 65.

The resultant voltage which is induced in the loop is 90° out of phase with that of the sense antenna, and leads or lags according to which edge of the loop coil is nearer the signal source. In compass operation, therefore, the direction in which the motor drives the loop coil depends on whether the signal leads or lags the sense signal.

The loop coil always rotates so that the same side of the coil is always toward the station being received. Since the loop coil position is transferred to the indicators by the autosyn system, the indicators show the bearing of the receiving station.

LOOP SET FOR NULL POSITION

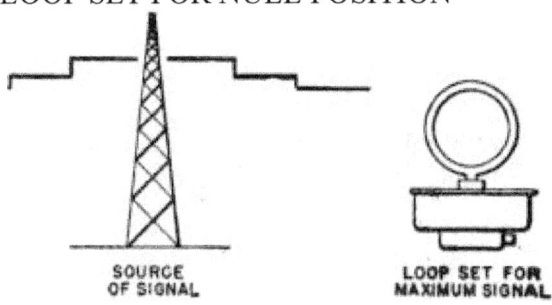

SOURCE
OF SIGNAL

LOOP SET FOR
MAXIMUM SIGNAL

Figure 65.—Loop

positions—Maximum and zero or null

signal.

QUIZ

1. Today the direct-reading magnetic compass is used as a

2. What is built into the compass to provide for expansion and contraction of the liquid resulting from altitude and temperature changes?

3. The purpose of the liquid In a compass is to of the card.

4. Top-reading compasses are usually graduated in degree steps.

5. The screws and brackets used for mounting magnetic compasses must be made of

6. are used in magnetic compasses to correct for deviation

errors.

7. Generally speaking, remote-indicating compasses operate on current.

a. Battery.

b. Pulsating.

c. Direct

d. Alternating.

8. The magnesyn compass consists of a and one or more

9. The gyro flux gate compass consists of , ,

and and also, when required, an inverter.

10. The caging knob is turned whether caging or uncaging.

11. Direct-reading magnetic compasses have an expansion chamber to provide for

12. Install compasses only where

13. The maximum allowable deviation of a compass before compensation is degrees.

a. 3.

b. 6.

c. 9.

d. 12.

14. Before swinging a compass, be sure the Liquid is .

15. If the compass reading is too low, the deviation is

16. The coefficient of north-south deviation is called coefficient

17. Compensation for coefficient "A" is done by

a. Moving the instrument in its mounting.

b. Moving the needle to the left.

c. Subtracting the deviation.

d. None of these.

18. In order to provide sufficient power to turn the induction motor

274033°—54 10 137

in the gyro-stabilized flux gate compass indicator, a(n)
is included in the system.

19. In a radio compass-loop the resultant voltage is degrees out
of phase with the sense antenna.

20. Direct-reading magnetic compasses are mounted two general ways: and

21. After compensation the maximum allowable deviation of a compass is degrees.

a. 2.

b. 3.

c. 4.

d. 5. ,

22. Before swinging the compass rose be sure that all magnetic equipment is

23. The directions on a compass rose on an air station are .

a. True.

b. Magnetic.

c. Agonic.

d. Isogonic.

24. If the compass reading is too high, the deviation is .

25. The coefficient of east-west deviation is knows as Coefficient

26. The remote-indicating magnetic compasses found in Naval aircraft are of the type.

27. The compass has its transmitting element stabilized.

28. The master indicator of a flux gate compass serves as a secondary
transmitter to send its readings to as many as secondary
indicators.

GYROSCOPES GYROSCOPIC PRINCIPLES

Before the development of aircraft instruments, all flight was limited to contact
conditions. The early pilot flew mainly by aligning his airplane with the horizon. By keeping his
wings parallel to the horizon he maintained level flight; by selecting some point on the nose of
the airplane and aligning it constantly with the horizon, he maintained a constant altitude.
However, when the weather closed in he "lost" the horizon. If the pilot flew at all when visibility
was poor, it was a hazardous undertaking indeed.

To make flight in adverse weather conditions possible, a reliable reference was needed—
within the airplane—which could be substituted for the horizon. Gyroscopic instruments were
developed to meet that need. They have been improved and perfected in design to match the
rapid improvement in design and performance of modern aircraft.

A gyroscope is a spinning wheel or rotor, which is universally mounted, that is, mounted
so it can assume any position in space. The model gyroscope shown in figure 66 portrays a rotor
that is free to spin about axis X-X on bearings in the inner ring or gimbal. The inner gimbal is
free to turn about axis Y-Y on pivots in the outer gimbal. The outer gimbal is free to turn about
axis Z-Z on pivots in the support. However, the illustrative model is a gyroscope in name only; it
does not exhibit gyroscopic properties until the rotor is spinning. To understand the functioning
of any gyroscope, it is necessary to understand two fundamental properties of an operating gyro.

Figure 66.—Elements of a gyroscope.

When the rotor of a gyroscope is spinning, its axis of rotation tends to remain in a fixed
direction in space. The rotor actually resists any force which attempts to change the direction of
its axis. Since the axis is at right angles to the plane of rotation of the rotor, any change in
orientation of the plane of the rotor is resisted. This property of an operating gyro is called

gyroscopic inertia, or, more simply "rigidity." In figure 07 the axis of rotation of the spinning rotor is horizontal. If the support is tilted, the inner gim-bal, in which the rotor spins, remains horizontal. If the support is swung in an arc. the spin axis continues to point in the same direction.

An operating gyro will resist a force which attempts to change the direction of its spin axis, but it will move in response to such a force or pressure. The movement is not a direct one in response to the force; it is a resultant movement. The gyro axis will be displaced, not in the direction of the applied force, but in a direction at right angles to the applied force, and in such a way as to tend to cause the direction of the rotation of the rotor to assume the direction of the torque resulting from the applied force. This property of an op-

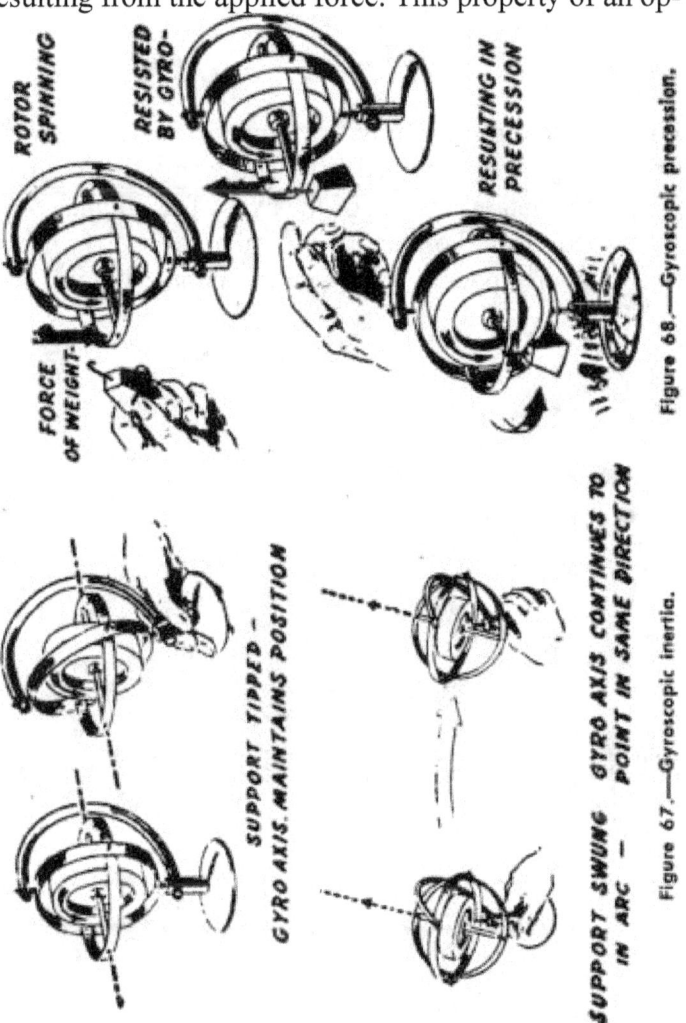

ROTOR SPINNING

RESISTED BY GYRO-

RESULTING IN PRECESSION

FORCE OF WEIGHT.

Figure 68.—Gyroscopic precession.

SUPPORT TIPPED - GYRO AXIS. MAINTAINS POSITION

SUPPORT SWUNG IN ARC —

GYRO AXIS CONTINUES TO POINT IN SAME DIRECTION

Figure 67.—Gyroscopic inertia.

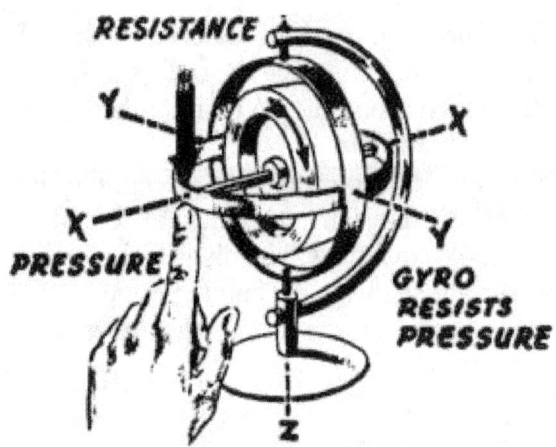

Figure 69.—Force applied to a gyro.

erating gyro is called gyroscopic precession, as shown in figure 68.

The forces acting on a gyroscope may be represented graphically as acting on the rotor itself, spinning freely in space, with a plane containing each of the axes of freedom.

For example, when a force is applied upward on the inner gimbal, as shown in figure 69, the force may be visualized as applied in an arc about axis Y-Y until it contacts the

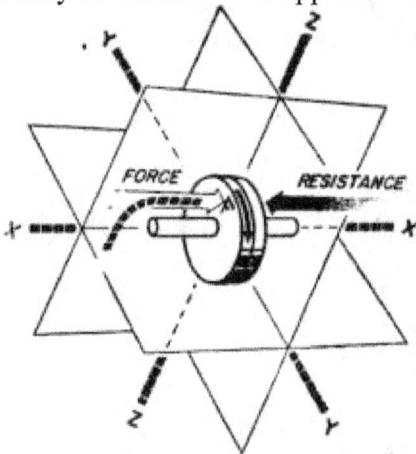

Figure 70.—Transmission of force. 142

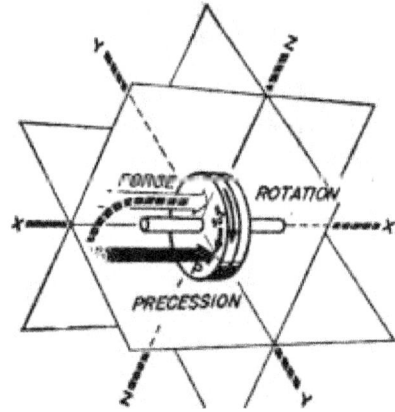

Flyurt 71 « — Direction of procottion.

rim of the rotor at F, as in figure 70. The effect produced by the force is equivalent to that produced by a force applied upward to the inner gimbal. The force at F is opposed by the resistance of gyroscopic inertia, preventing the rotor from being displaced about axis Y-Y. With the rotor spinning clockwise, the precession will take place 90° ahead in the direction of rotation at P, as in figure 71. The rotor turns about axis Z-Z in the direction of the arrow at P, as in figure 72.

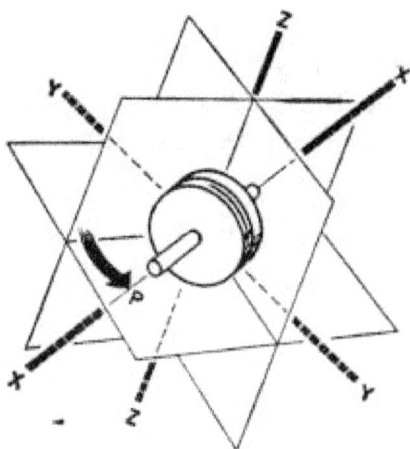

Figure 72.—Processional movement 143

These two fundamental properties are utilized in gyroscopic instruments. Gyroscopic inertia establishes a reference in space, unaffected by movement of the supporting body. Precession is utilized to control the effects of drift, whether it is apparent drift or mechanical drift, and maintains the reference in the required position.

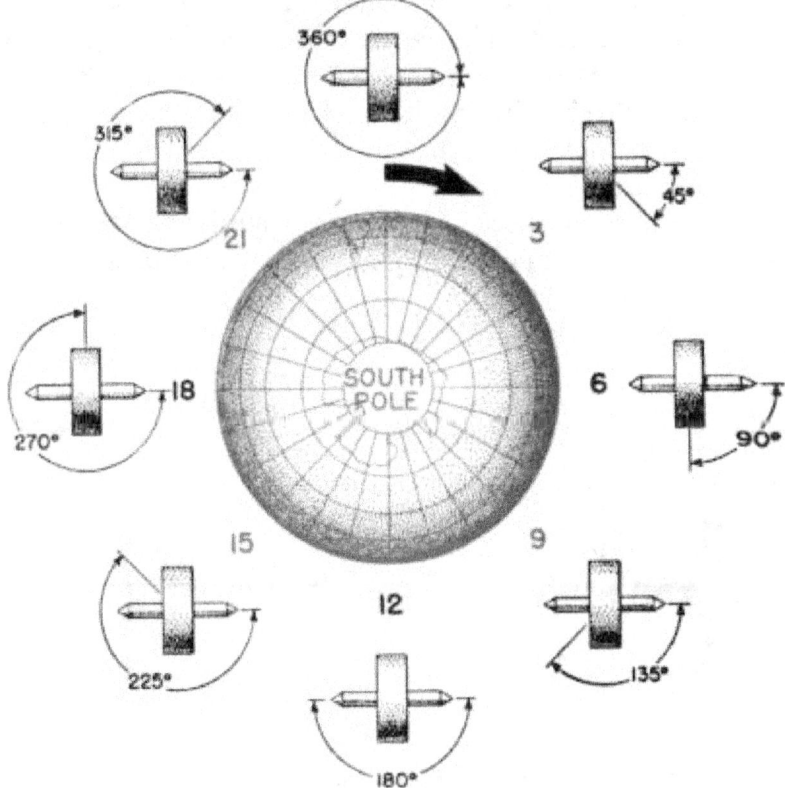

Figure 73.—Action of a free gyro.

A free gyro (one not provided with an erection system) maintains its axis fixed in relation to space, and not in relation to the surface of the earth. For example, imagine such a gyro at the equator, its starting position being with the spin axis horizontal and pointed in an east-west direction. See figure 73. The earth turns in the direction of the arrow, or clockwise, with an angular velocity of one revolution every 24 hours. To an observer out in space, the spin axis would appear to maintain its direction pointing east.

However, to an observer on the earth, stationed where the gyro is, the spin axis appears gradually to tilt or drift. At the end of 3 hours the spin axis has tilted 45°; at the end of 6 hours, the spin axis has tilted 90° and is in a vertical position. At the end of 12 hours the spin axis is again horizontal but pointing west; at the end of 24 hours it is back where it started. See figure 74. This action of a free gyro is known

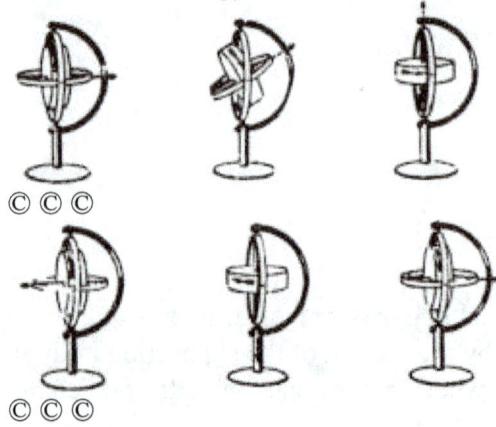

Figure 74.—Apparent drift.

as apparent drift. To overcome apparent drift as well as mechanical drift caused by bearing friction or slight unbalance, a gyroscopic instrument must be provided with an erecting device which maintains the spin axis in the required position. This erecting device applies a force to the gyro whenever drift occurs. Precession returns the spin axis to its normal position, maintaining an accurate reference, as shown in figure 75.

By using a gyroscope with an erecting mechanism within an airplane, a stabilized reference is provided. As the airplane rolls or pitches, the spin axis remains upright, as shown

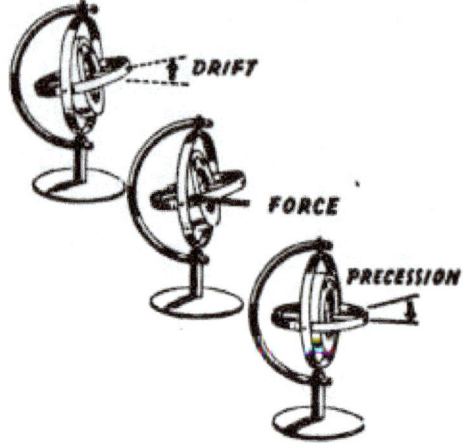

Figure 75» Pr©C6ttlon controlling.

in figure 76. When an appropriate indicating device is attached to the gyro, deviations from the norrrlal attitude of the airplane will be shown. By referring to the indications, the pilot may control the airplane.

GYRO OPERATION IN INSTRUMENTS

In the early days of aircraft instruments, those being gyro operated had their gyros driven by a stream of air. The

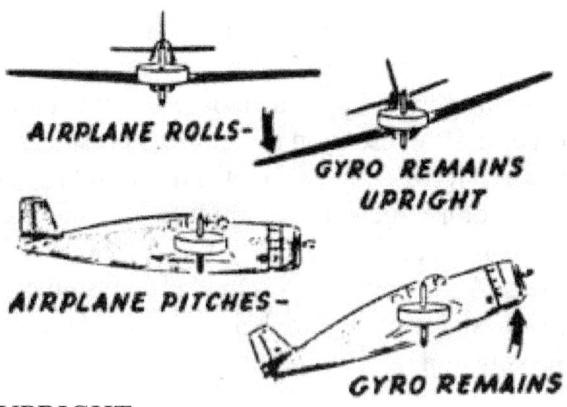

AIRPLANE ROLLS-

GYRO REMAINS
UPRIGHT

AIRPLANE PITCHES-

GYRO REMAINS

UPRIGHT

Figure 76.—Aircraft reference. 146

airstream was caused by suction developed either by a venturi tube or by an engine-driven vacuum pump. The venturi tube could obtain a vacuum from the flow of air passing the aircraft only during flight. It operated according to a law

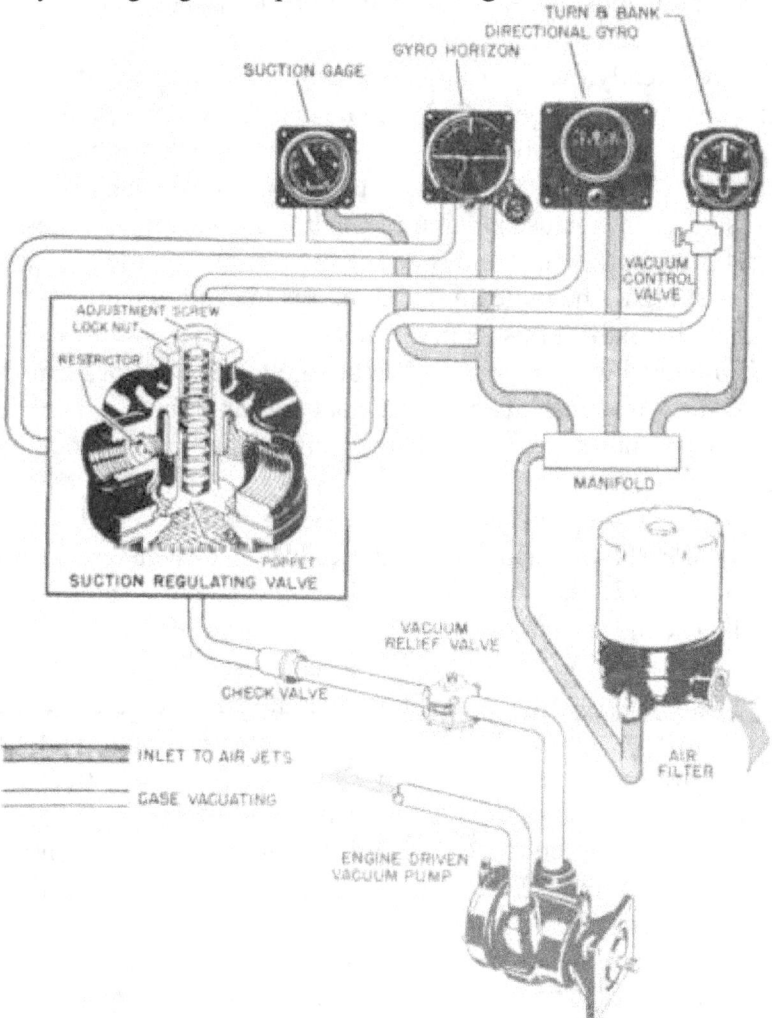

Figure 77.—Vacuum system showing cutaway of suction regulating valve.

of physics that pressure decreases as speed increases along a streamlined restriction. The engine-driven vacuum pump was designed to augment and later replaced the venturi tube. Figure 77 illustrates the vacuum pump method of instrument operation and shows the suction gage used to indicate the amount of suction provided for gyro instruments. An engine-driven vacuum pump would provide a better, more stable, and reliable source of vacuum. However in modern aircraft, these gyro driven instruments are being operated by an electric source. An electric-driven gyro rotor is used that is quite suitable for high altitude flying as compared to the air-driven gyro rotor which at high altitudes may encounter inadequate air supply because of low atmospheric pressure.

This description of the change in methods of gyro operation is made because there are still cases in which air-driven gyros are used. But as new planes are being developed and released the air-driven gyros are becoming obsolete #nd are being replaced by electric-driven gyros.

TURN AND BANK INDICATORS

The turn and bank indicator is installed in aircraft to indicate the lateral attitude of the aircraft in straight flight and to provide a reference for the proper execution of a coordinated bank and turn. It also indicates the flying of a straight course or the direction and rate of a turn. It is put to its most important use, and is most necessary, when flight conditions of poor visibility are encountered.

A turn and bank indicator is really two instruments mounted as a single unit on the instrument panel. The turn indicator part is a gyroscopic device that shows, by movement of its indicator hand, the direction in which a turn is being made. A slow turn to the right causes the hand to move over the dial a small distance to the right. A fast turn makes the hand move considerably farther. Left turns, of course, cause hand movement to the left. In all cases, the distance of the hand movement is proportional to the speed of the turn. When the airplane finishes its turn and again heads straight, the indicator hand returns to center position.

The bank indicator—the other half of this team—is not

a gyro instrument. It is mounted with the turn indicator solely for the pilot's convenience, since information on banks and turns is almost always wanted at the same time. A bank indicator is sometimes called an inclinometer, and is a very simple mechanism that looks something like a carpenter's spirit level.

The main part of a bank indicator is a glass ball which moves in a curved glass tube filled with liquid. The tube is mounted horizontally across the front of the instrument dial, a little below the center, as you see in figure 77. The liquid in the tube is for the purpose of damping the. movement of the glass ball. The movement of the glass ball is controlled by two forces—(1) gravitational pull of the earth, or gravity, and (2) centrifugal force caused by the turning of the aircraft. When the airplane is making a properly banked turn, the ball stays in its center position because of the proper balance of these forces. If the ball moves in the direction of the turn (low side of the tube), it indicates that the aircraft is slipping toward the inside of the turn as a result of overbanking. Therefore, the force of gravity controls the movement of the ball.

If the ball moves in the opposite direction (to the high side of the tube), you know that the aircraft is skidding toward the outside of the turn as a result of underbanking. Therefore, centrifugal force controls the movement of the ball.

The main element in the turn indicator is a small gyro rotor, mounted in a frame that is pivoted to turn on a longitudinal (fore-and-aft) axis. The rotor is kept spinning by a stream of air from an intake jet on the instrument case or by a 24 to 28 volt, direct-current motor mounted in a gimbal which serves as the sensitive element or gyro for the turn indicator. In the air-driven type air is withdrawn from the case through a connection to the pump-operated vacuum system of the airplane, causing air from the cockpit to enter through the intake jet. The air stream is directed against the bucket of the rotor, causing the gyro to rotate at high speed. The rotor is carefully balanced, and runs on pre-

cision ball bearings which are supplied with oil from a reservoir inside the gyro.

The air- and electric-driven gyros operate similarly and on the same principles except for source of power. The air-driven type will operate only when the aircraft engine is running.

Because of the way in which its frame is mounted, the gyro responds only to turns around a vertical axis—that is, turns to the right or left. It is not affected by rolling or pitching of the airplane. But what happens during a turn? The gyro goes into its precession act. Refer to figure 78 as you read the explanation.

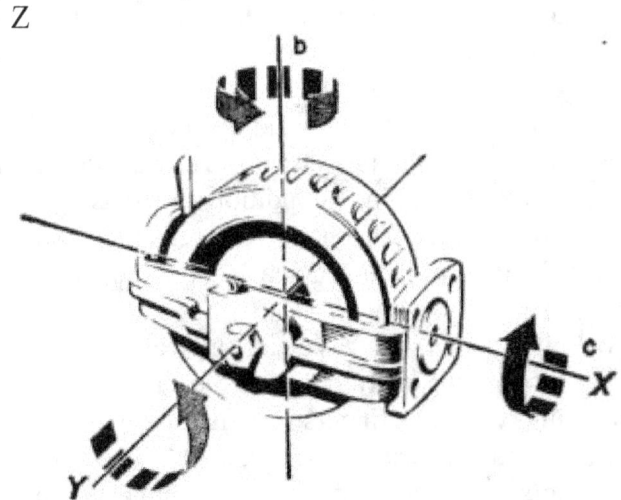

Figure 78• Gyro rotor and frame, bonk and turn indicator.

The rotor spins on the Y axis, as you see in the illustration. If, for example, the airplane turns to the left, the gyro assembly is rotated on the vertical or Z axis. The reaction of the gyro to this turning influence is an immediate rotation (in the direction of arrow C) about the X axis, which is at right angles to the axis of the airplane turn. You'll understand this if you remember how precession takes place. To refresh

your memory, precession is the natural reaction at right angles to a twist applied to a gyro, and is a fundamental gyro characteristic.

In the turn indicator, the rotation of the gyro assembly around the X axis is partially held in check by a spring, and is limited by stops from going beyond approximately 45° past the vertical in either direction. The spring helps to balance the gyro reaction during a turn, and brings the assembly to its neutral vertical position as soon as the airplane returns to a straight flight. The action of the gyro •during a turn is also damped or smoothed out by a plunger arrangement known as a dash-pot. The combined efforts of the gyro, the spring and the dash-pot damper are balanced so as to make the movement of the indicator hand proportional to the rate at which the airplane turns.

The instrument case of the air-driven type is fitted with connections for cutting in the vacuum system, a damping mechanism adjustment, a sensitivity adjustment, and a lubrication opening. One of the vacuum system connections is to the rear on the bottom of the case; the other is on the case back. The two connections are provided for convenience in installation and only one is used at a time—the other being covered by means of a pipe plug.

On the electric-driven type, the rear plate assembly of the instrument case contains the electrical receptacle.

On the air-driven type the damping adjustment screw and lock nut are on the right side of the case, just behind the mounting flange. When this screw is turned "in," the open area of the damping orifice is increased, lessening the damping effect. The screw for adjusting the sensitivity of the turn indicator also has a lock nut, and is located on the left side of the case behind the mounting flange. Turning the sensitivity screw "in" decreases the tension on the check spring.

The turn and bank indicator is usually grouped with the other flight instruments on the airplane instrument panel. It should be mounted so that the dial is vertical when your plane is in normal flight position. And be sure the ball in

the bank indicator tube rests in center position when the airplane is level.

Under normal flying conditions a vacuum of about 2.0 inches Hg. is correct inside the case of the instrument. You should check the vacuum after installation by connecting a suction gage to the case connection which is not being used. Be certain that the pipe plug is replaced on this connection after the test has been made. Varying temperatures and characteristics of individual instruments may mean loss of sensitivity in some instruments. The vacuum should be set to produce proper deflection of the pointer under prevailing* conditions. Increasing the vacuum (by screwing the valve stem outward) makes the instrument more sensitive, and decreasing the vacuum makes it less sensitive. Under no circumstances should the vacuum be less than 1.80 inches or more than 2.20 inches Hg.

The screen over the jet by which cockpit air enters the cas.e can be cleaned by removing the jet assembly and washing it in a cleaning solvent #PS-661a or a suitable substitute. When it's thoroughly dry reinsert the assembly. Tightening with your fingers is sufficient.

No operating instructions are required for the electrical type. Three minutes after power is applied, the instrument should operate continuously and without further attention. When installing the instrument, the mounting flange of the indicator must be grounded to the structure of the aircraft by bonding to the instrument panel. No lubrication is necessary except at overhaul periods. If the pointer fails to respond (gyro does not operate) there is probably a poor electrical connection, the operating voltage is too low, or there is an open circuit in the instrument. To remedy this, the connection should be checked, rated voltage should be supplied, or the instrument should be removed and sent to overhaul. Should the pointer indicate turn in the wrong direction, the wires are reversed on the instrument. The wires should be disconnected and reversed. If the pointer is off zero, the motor assembly is unbalanced and the instrument should be removed for overhaul.

DIRECTIONAL GYROS

Flying an airplane would be a lot easier if there were traffic cops and signposts in the sky. The directional gyro is one of several instruments that helps to make up for this lack by providing a "reference point" when you're trying to keep on course or make an accurate turn to a new heading. ,, This gyro instrument can do such a job because it obeys the basic gyroscopic principle of rigidity in space. It supplements the compass in following a fixed course. It shows the exact numbe°r of degrees turned when you change the airplane heading. It helps maintain alinement when you have to make an instrument landing. It's of great value in helping the pilot and the navigator to check each other's calculations, It assists in locating the direction of radio beacon stations. And it coordinates the problems of bombardier and pilot on tactical missions.

The rotor of a directional gyro is mounted on a gimbal ring, and is free to turn around the vertical as well as the horizontal axis of the ring while spinning around its own horizontal axis. This is called a "universal" mounting. A caging knob on the front of the instrument permits you to turn the whole mechanism—the gimbal ring, the rotor, and the reference card. This card, which is read through a rectangular glass panel on the front of the instrument, is marked like a pilot's compass—in degrees—and is sometimes called the "azimuth" card.

Some directional gyros have air-driven rotors while others are driven by electrical power. Both work on the same principles of gyroscopic action.

In the air-driven gyro instrument, the rotor is turned by an air stream. Air is withdrawn from the instrument case by the vacuum system of the airplane. Cockpit air pressure thereupon forces a new supply of air into the case through a screened opening on the case. The air stream is directed through a jet. The air from the jet is then directed through a nozzle, onto the buckets in the rim of the rotor.

When the rotor is upright, air strikes the rotor buckets squarely in the middle. If the rotor tilts, air from the jet

strikes on one side of the bucket instead of striking squarely causing the rotor to return to its upright position.

Don't get the idea that a directional gyro is a type of compass. It's not in itself north-seeking, but must be set to correspond to the compass heading or to whatever heading you desire as a reference. Under average flying conditions the gyro will stay in its fixed position in space so that its rotor axis points constantly in one direction. The card, mounted on the vertical ring and turning with it on the vertical axis, shows any deviation of the airplane from its course.

A certain amount of error or drift, caused by* bearing friction and certain other minor factors, is present in the operation of the directional gyro. This error is quite small, and frequently runs to only about 3° in a period of 15 minutes. The gyro should be checked against the magnetic compass at frequent intervals if it's to be of most use to you as an accurate "guidepost."

The instrument is built so that it may be tipped about 55° from normal in any direction. But if the airplane pitches or banks more than 55°, the gyro will be upset. The card will probably start to spin, and you won't be able to make head or tail of its readings until the airplane returns to level flight and the gyro is caged and reset. Always cage the gyro before doing airplane acrobatics. This will prevent damage and prolong the instrument's life.

When resetting a directional gyro, be sure that a correct reading is obtained from the magnetic compass, which usually oscillates considerably in rough air—even when the airplane is flying a straight course. When the card of the magnetic compass appears to be still it is probably at the end of a swing, and therefore considerably off the heading. Watch the compass for about a minute while the plane is held on a straight course, and note the average reading between swings in opposite directions.

WTien you're installing a directional gyro, be sure the instrument is leveled properly. If necessary, you can add shims between the turn indicator and the panel at top or bottom.

Figure 79.—Directional gyro dial.

but don't use shims unless they're of uniform thickness. In other words, be careful that the shims don't twist the aline-ment of the instrument to right or left.

After the instrument is installed, the vacuum supply to it should be checked. Four inches Hg. is the ideal pressure reduction to operate the rotor. Two other tests—the coast test (on oil-lubricated instruments) and the drift test—are made, to check on proper operation. In the coast test, the gyro is run for 5 minutes or longer on the correct amount of vacuum, after which the engine is shut off. After 8 minutes you should check to see if the gyro is still spinning. An oil-lubricated gyro that won't coast for more than 8 minutes is considered unserviceable. New models of directional gyros are being produced which are lubricated with grease. They will not coast for more than 3 or 4 minutes even when in good condition, so the coast test doesn't apply to them.

The drift test is made by running the gyro for 10 minutes on each of the cardinal headings (0°, 90°, 180°, and 270°) with the airplane in tail-down position. The card should not drift more than 5° on any one cardinal heading and no

more than 3° on any of the other cardinal headings, and the total drift for all four cardinal headings should not exceed 12°.

GYRO HORIZON INDICATORS

A pilot instinctively checks the position in which his airplane is ftying by referring to the horizon—if and when he can see it. Often enough, however, the horizon isn't visible. Darkness, storms, clouds, fogs, and other weather conditions may hide it Even an old-line pilot who "flies by the seat of his pants" is bothered plenty when the horizon disappears.

Pilots have a substitute for the earth's horizon in the form of a horizon indicator on the instrument panel in the cockpit. You've probably heard horizon indicators called by several names, such as flight indicators, gyro horizons, artificial horizons or bank-and-climb indicators. They all refer to the same kind of instrument and do the same job. From them you learn the relative position of your airplane with reference to the true horizon.

The horizon indicator has a small gyro rotor assembly set so the flywheel spins in a horizontal plane. A bar on the

Figure 80.—Dial of a horizon indicator.

dial face of the instrument indicates fore-and-aft inclination of the airplane over a range of 60° in either an upward or a downward direction. A pointer on the same dial face also shows lateral inclination (or "bank*') up to 90° toward

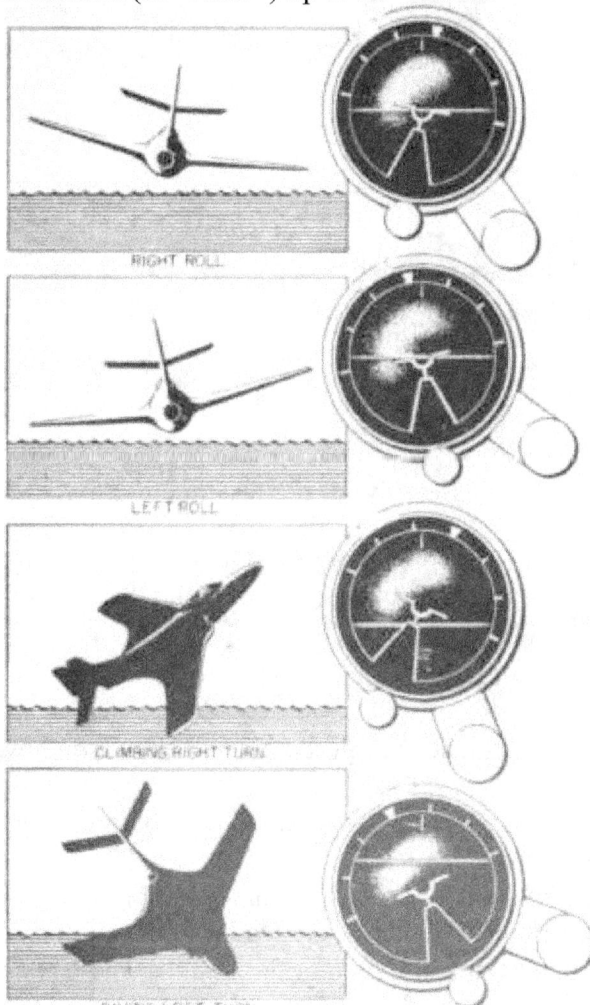

Figure 81.—Roll and pitch indications.

*

either side. A dial is seen in figure SO, while roll and pitch indications are shown in figure 81.

The case of the gyro horizon indicator is rigidly attached

to the level of the aircraft, therefore, any movement of the airplane is synchronously and identically duplicated by the case. The case is free to revolve around the stable gyro because of the mounting of the gyro motor in gimbals. It follows, therefore, that the airplane, itself, actually revolves around the rotor and is the complementing factor in estab-

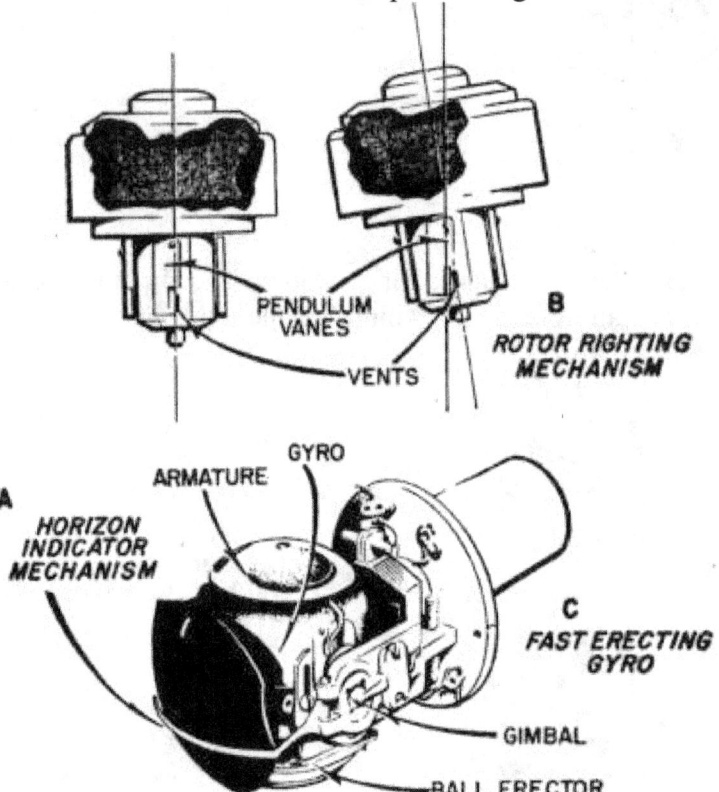

Figure 82.—(Al Horizon indicator mechanism; (B) rotor righting mechanism, showing jots and pendulum vanes; and (C) gyro and gimbal assembly off the fast erecting gyro horizon indicator.

lishing the indications of the instrument, even though, to the observer, this movement seems to be the reverse. The understanding of these facts leads to a better conception of the principles governing gyroscopic instrumentation.

The gyro assembly is equipped with a righting mechanism which goes into action whenever the rotor axle is tipped

away from the normal vertical position. A glance at figure 82B will show you how this mechanism works. A righting torque—or twist—is provided by air jets operating as part of the vacuUm system. Four pendulum-like vanes are suspended around a cylinder below the gyro housing. Each of these vanes partially covers one of the four air vents (see fig. 82B) that exhaust the air from the gyro compartment. As the gyro starts to tip from its axle-upright position, the pull of gravity holds the vanes vertical. The tipping of the gyro causes one vane to close one vent, but the opposite vane completely opens its vent, as shown in figure 82B. The reaction of the air from this open port causes the gyro to precess back to its normal position.

A look at figure 82C shows a view of the electric-driven gyro type indicator. It is known as the fast erecting gyro horizon indicator, which was made necessary by the development of jet-propelled aircraft where the engine warmup time is comparatively short, and where it is necessary that the gyro be brought up to the erected position in a very short time after the power is turned on. Its operation is like that described above except its power source is electric instead of air.

The horizon indicator has no time lag, and is easy to read after a short familiarizing period. Like all gyroscopic instruments, it is precision-made and therefore must be treated with care. It should be mounted on a vibration proof instrument panel and attached by screws through the mounting lugs. Be sure the instrument is leveled properly after installation. If its face is not exactly vertical, shims should be added between the lugs and the front of the panel until the horizon bar lines up with its index markers at the sides of the dial. When you are leveling or alining, the aircraft must, of course, be in flying position (tail up, fuselage level).

If the gyro horizon indicator is properly installed, it should not require overhaul before 1200 hours of operation. The instrument should be removed from the plane and sent to instrument overhaul for repair. No lubrication is required between overhauls. Most troubles encountered will probably

be due to improper power supply (electrical or air) or to excessive vibration. Normally, if the indicator fails to respond, the probable cause is failure of the power supply, faulty power supply connections, or faulty internal wiring' of mechanism. If the indicator and horizon bar do not settle properly or respond when the aircraft maneuvers, most likely the indicator will need replacing.

QUIZ

1. When the rotor of a gyroscope is spinning, its axis of rotation tends to

2. An operating gyro will a force which attempts to change
the direction of its spin axis.

3. When an outside force acts upon a gyro, it will precess
a. 90 degrees ahead of the direction of rotation.
b. 90 degrees behind the direction of rotation.
c. In the direction of the force.
d. Against the direction of the force.

4. To cause a gyro to maintain itself in relation to the earth, a(n) device is installed on the gyro.

5. Aircraft instrument gyros are normally driven by

6. The is used to obtain vacuum from the flow of air
passing the aircraft during flight.

7. Lateral attitude of an aircraft in flight is indicated in the aircraft by the indicator.

8. Vacuum in a turn and bank indicator under normal flight conditions should be about inches.

9. In the airdriven gyro instrument, the rotor is driven by
air from the case.

10. An oil lubricated gyro that won't coast for at least
minutes is considered unserviceable.

11. When a gyro is displaced it will move to the applied force.

12. The turn-and-bank indicator indicates of the aircraft.

13. The part of the turn-and-bank indicator is not a gyro

instrument.

14. The action of the turn-and-bank gyro is damped during a turn by a plunger arrapgement known as a

15. With the electric turn-and-bank indicator, if the pointer fails to respond, the defect is probably

16. Always the directional gyro before doing acrobatics.

17. The horizon indicator has time lag.

a. A 3 second.

b. A 2 minute.

c. A short.

d. No.

18. The gyro horizon indicator should not require overhaul before hours of operation.

a. 2000.

b. 500.

c. 1200.

d. 1800.

19. A gyroscope is a spinning wheel or rotor Which is mounted.

20. A free gyro maintains its axis fixed in relation .

a. To the earth's surface.

b. To space.

c. To the north pole.

d. To the deck of the airplane.

21. The venturi tube is used to obtain vacuum from the

22. In all cases, the distance the hand of a turn-and-bank indicator moves is to the of the turn.

23. The bank indicator is sometimes called

24. The turn-and-bank indicator is usually grouped with the instruments.

a. Control.

b. Flight.

c. Engine.

d. Navigation.

25. The directional gyro may be tipped about degrees from normal in any direction without upsetting.

26. inches of mercury is the ideal pressure reduction to operate directional gyro rotor.

a. Two.

b. Four.

c. Six.

d. Eight.

27. The gyro horizon indicator has a small gyro that spins in a plane.

28. The gyro horizon indicator requires .

a. No lubrication at all.

b. No lubrication between overhauls.

c. Lubrication at every 120 hour check.

d. Lubrication every check period.

CHAPTER
AUTOMATIC PILOTS

The automatic pilot is a wonderful instrument. Because it is able to do several remarkable things, some people think it must be very complicated and difficult to understand. That isn't so. True, the automatic pilot has a fairly large number of parts, and takes longer to install than most instruments, but its complexities have been greatly overrated.

The purposes of the automatic pilot are to relieve the human pilot while he takes care of other duties, and to keep the airplane on an even, straight course more exactly than human hands could do on bombing runs and the like. Of course, in addition to maintaining mechanical control over the airplane, the automatic pilot also gives instrument-panel readings just like a directional gyro and a horizon indicator combined. And while it's used mostly to maintain straight, level flight, all normal maneuvers can be performed by manipulation of the instrument's simple control knobs, and with more precision than is possible by a human pilot.

TYPE S-3 AUTOMATIC PILOT

The purpose of the S-3 hydraulic automatic pilot is to control the movement of the airplane about its rudder, aileron, and elevator axes. The gyro in the directional gyro control unit furnishes the fixed reference for any movement about the rudder axis. Any deviation from a straight course is shown visually by the directional gyro card. At the same time the movements of the gyro case about the gyro rotor causes the servo unit (hydraulic surface control) to act on the rudder, bringing the airplane back to its heading. The gyro in the bank and climb unit furnishes the

fixed reference for any movement about either the aileron or elevator axis. Any roll or pitch is shown visibly by the horizon bar. Movement of the gyro case about the gyro rotor causes the aileron or elevator servo unit (hydraulic surface control) to correct the deviation.

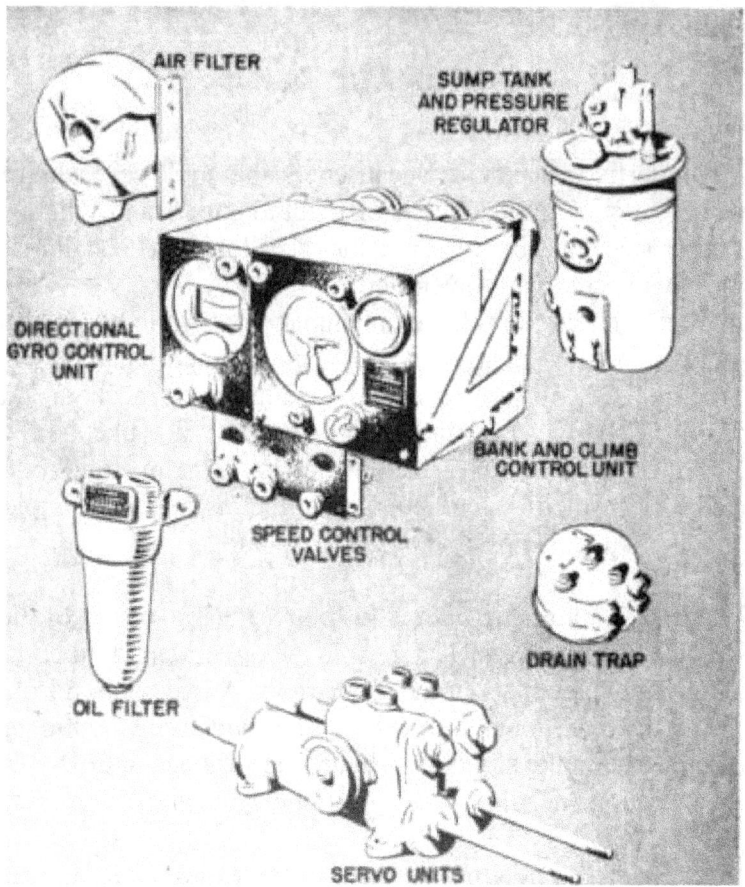

Figure 83.—Composite view of parts of S-3 automatic pilot.

The S-3 automatic pilot consists essentially of two gyro control units, a mounting unit, servo units (hydraulic surface controls), speed control valve, air filter, oil filter, combination sump tank and pressure regulator, drain trap (if

required), and manifold block. Other accessories include an oil pressure gage, a vacuum regulating valve, a vacuum pump, an oil pump, and an oil separator.

The S-3 automatic pilot makes automatic control possible so that the aircraft will maintain the heading and attitude fixed by the human pilot. It provides visual indications of the attitude of the aircraft in yaw, pitch, and roll. A composite view of all parts of the S-S automat ic pilot is shown in figure 83. Figure 84 shows the cable connections of a typical installation.

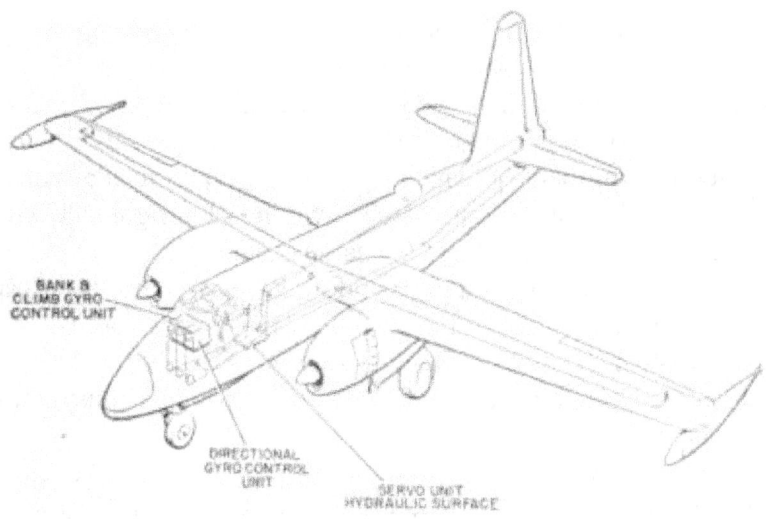

CONTROL

Figure 84.—Schematic of cable connections of a typical installation.

The S-3 automatic pilot operates on a system which combines two gyroscopes with pneumatic and hydraulic mechanism to work the controls in coordination. One of the gyros is called the directional unit and is quite similar to a regular directional gyro in that it supplies the reference for directional control (rudder). The other gyro is the bank-and-climb unit, which provides the reference for both lateral (aileron) and longitudinal (elevator) control. The two units are illustrated in figure 85.

In talking about the operation of this model of automatic

Figure 85.—Bank and climb unit and directional unit.

pilot, only the aileron control of the bank-and-climb control unit will be discussed, as the rudder and elevator controls operate in the same way.

ELEVATOR KNOB

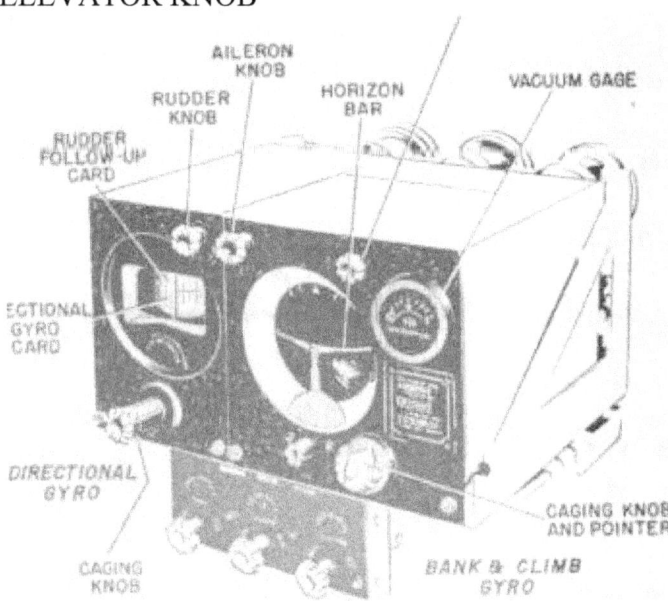

Figure 86.—Gyro control units.

The bank and climb gyro spins with its axis vertical, like a horizon indicator, and is kept erect by the same type of vane-and-vent mechanism. Looking at figures 86, 87, and 88 as you follow through this description will help to clarify nomenclature involved.

With the airplane flying level, the horizon bar on the face of the bank and climb unit is level, and the miniature airplane is parallel to it. The gyro remains fixed in position due to rigidity as the airplane banks, and the degree of movement is indicated on the dial, as you already know from your study of the horizon indicator.

The gyro rotor is supported in a gimbal ring which has a disk with knife edges attached to it. Those parts, which are called the "sensitive element," are shown in red in figures 87 and 88. The air "pick-offs" (A-A' of figs. 87, 88) are enclosed in the same casing as the gyro element. Air is drawn into the back of the casing by the suction pump B and directed onto the gyro to spin it. The pump also draws in air from the relay C through the pickoff or nozzle ports A-A'.

The air relay C has two inlet ports (D and E) on either side of its diaphragm F. This diaphragm is connected by a piston rod G to a balanced oil valve H, in which a constant oil pressure is maintained by the oil pump J.

When the core of the balanced oil valve moves to left or right, it permits oil to flow to the servo unit K, where it moves the piston rod L-L/ one way or the other. The piston rod is connected to the control cables which operate the ailerons.

In figure 87 the aircraft is flying level, so the automatic pilot is neutral. The gyro wheel is upright and the knife edges of the disk intercept an equal amount of air being drawn in from the air relay at D and E. In level position, an equal vacuum is maintained at both sides of the diaphragm F, the oil valve piston is centered, and no oil can flow to the servo cylinder K. The pressure regulator N maintains constant pressure at the balanced oil valve and permits the oil to flow back to the sump O.

Now take a look at figure 88, which shows that the aircraft has tilted to the right so that the right wing is lower than

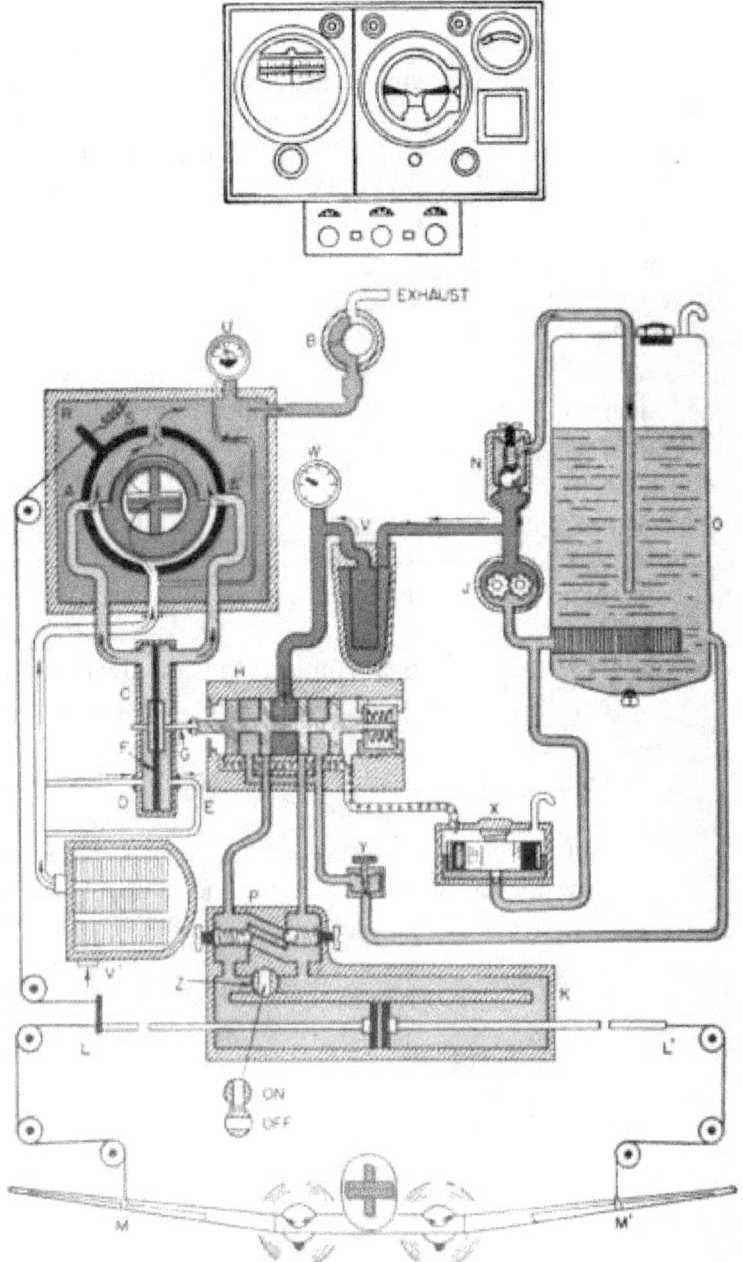

Figure 87.—Diagram of aileron control, level flight.

OB DARK GREEN = PRESSURE FLOW
■B RED = GYRO ELEMENT
CZ3 LIGHT GREEN = EVACUATED AREA
Ml DARK BLUE = OIL UNDER PRESSURE
EZ3 LIGHT BLUE - OIL AT ATMOSPHERIC PRESSURE
Figure 88.—Diagram of aileron control, right wing low.

274033°—54 12 169

the left. The gyro (shown in red) stays vertical and thus closes port A' of the air-pickoff system. At the same time port A is opened, the suction on the left side of the diaphragm F is increased. Thus, the diaphragm moves to the left and pulls the piston of the oil valve with it. The movement of the valve lets oil flow through pipe number 1 to the servo unit, where it passes around the overpower valve P and enters the servo unit cylinder. The oil pressure moves the servo piston to the right, and thus applies the necessary aileron control to bring the aircraft back to level flight. The oil from the other side of the piston returns through the balanced oil valve to the sump.

Another important part of the system is the followup, which also is shown in figures 87 and 88. The followup gradually removes the aileron, control applied by the servo system, and goes into action while the airplane is returning to normal position so the control surface will be back in neutral position when the airplane is restored to level flight Actually, the air pickoffs A-A' are not fastened rigidly to the gyro box, but can be moved in relation to the gyro by means of the followup mechanism, or by turning the trim knobs.

A cable is connected to the servo piston rod and attached to the lever R (top left of fig. 88) on the followup assembly. The movement of the followup is not shown in figure 88 because it would change the position of the air pickoffs, but the following action would take place:

When the servo piston L-L' moves to the right, the followup cable moves in the same direction and rotates the followup assembly against the pull of the balance spring S. This moves pickoff A down and A' up. When these ports reach a neutral position (both halfway open), the air relay and the oil valve are centered. When this occurs, movement of the servo piston away from neutral ceases.

As the control surfaces (in this instance, the ailerons) continue to bring the airplane back to normal attitude, the air pickoffs pass beyond the neutral point and begin to cause servo movement in the opposite direction. Remember, this

is not opposite control being applied. It's merely the removal of the control originally applied.

The followup mechanism, as you read earlier, is made so that the correct amount of control will be applied, and later removed at the proper rate as the airplane returns to normal attitude of flight.

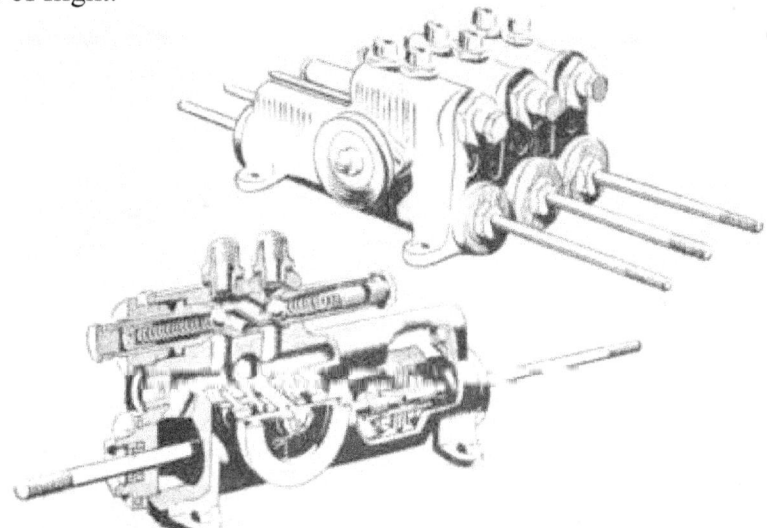

ngure oY.-^aervo unn nyaraunc surface control uniT.

Broadly speaking, that's all there is to the operation of the S-3 Automatic Pilot because the mechanisms which control the rudder and elevators work the same way. There are, of course, a number of accessories on automatic pilots, some of which you see in the diagrams. For example, there is a suction regulator which keeps the gyro-spinning vacuum—and that for the pickoffs—at the proper value regardless of the speed of the suction pump. The "vacuum supply" in inches of mercury is indicated on the vacuum gage U.

The oil sump O carries the oil supply. A filter at V prevents foreign matter from reaching the hydraulic system. N is a valve which automatically regulates the oil pressure from the pump and circulates it through the sump whenever

the balanced oil valve cuts off circulation to the servo unit. The oil pressure is indicated on the oil gage W. When the balanced oil valve is located below the level of the sump, a drain trap X is used in the installation to collect the drain oil from the balanced oil valve to the sump.

The servo relief valves, shown at P, allow the human pilot to overpower the automatic pilot when the system is in operation. Speed control valves, like the one shown at Y, regulate the oil flow from the servo pistons and thus govern the speed with which the automatic pilot operates the controls.

The bypass valve Z in the servo unit is used to turn the automatic pilot on or off. The valve is—in practice—connected by pulley and cable to a lever where the human pilot can reach it. When he wants to fly the airplane manually, he opens the valve. Oil then flows through the bypass tube from one end of the servo to the other, and the controls then can be moved freely because the pressure on both sides of the servo piston is equalized.

The directional gyro control unit of the automatic pilot contains the gyro which is the reference for automatic or manual rudder control. It also contains a ball-type bank indicator, the air pickoffs, and a device for setting the automatic pilot to hold any selected heading.

In order to have the automatic control guide your plane on a selected heading, the air pickoffs must be at neutral when the airplane is headed so as to show the selected reading on the directional gyro. Notice that this unit has a second card above the regular directional gyro card. Tho upper, or "followup" card is attached directly to the pickoffs. The pickoffs are at .neutral position when the readings on the follow-up and the directional gyro cards coincide.

The position of the followup card is controlled by the rudder knob above the instrument dial. The lower, or directional, card may be set to any heading desired by pushing in and turning the caging knob underneath the dial.

The bank and climb control unit contains a gyroscope on a vertical axis; this gyro controls the ailerons and elevators. The bank-and-climb control unit also contains air pickoffs

for the two controls, and median isms for setting the automatic pilot so it will fly the airplane in the desired attitude.' The unit also contains a differential vacuum gage to give the pressure drop across the rotor.

The.circular dial on the S-3 instrument is attached to the gimbal ring of the gyro. Thus, the dial provides you with a fixed horizontal reference as the airplane banks one way or the other. The amount of bank is shown on a scale on the top of the dial.

The horizontal bar in front of the dial is moved by linkage from a pin in the side of the gyro case. The bar rises as the airplane noses down, descends as the airplane noses up, and remains horizontal as the airplane banks. A miniature airplane is positioned in front of the horizon bar, and gives the pilot a visual picture of how his craft is positioned with respect to the earth. To compensate for variation in load conditions, the miniature airplane can be raised or lowered by means of a knob under the dial.

There's a pointer at the right-hand side of the horizon dial. It's an alinement index for lining up the elevator fol-lowup index when your airplane is flying level—preparatory to engaging the automatic pilot. There are also alinement indices for the ailerons. The aileron knob is for setting the desired angle of bank when you wish to maneuver the airplane by operating the automatic controls instead of the regular manual ones. It changes the position of the air pickoffs.

The elevator knob does the same thing for longitudinal control—setting the desired angle for climb or glide.

The two control units you have just been reading about are supported by a mounting unit into which they slide in place on tracks. The mounting unit is a frame to which the air relay balanced oil valves, followup mechanism pulleys, pressure oil and drain oil manifolds are attached. All the necessary electrical, mechanical, and air connections are automatically made when the control units are bolted in place.

For example, the followup pulleys, to which the followup cables are attached, are provided with clutches which carry their motion to the control units. The pressure and drain oil manifolds on the bottom of the mounting unit are piped to the three balanced oil valves. All the air intake connections, for both the air relays and gyros, open into a manifold, and the entire air system can thus be connected through a single air filter.'

Speed valves are made part of the servo oil system. They control the rate of flow of oil from each servo cylinder to the sump, and thus regulate the speed with which each servo responds. You'll find the valves connected in the return line from the balanced oil valves to the sump. An oil pressure gage in the cockpit shows the pressure at which oil is being supplied to the automatic pilot. The proper operating pressure for servo oil will differ with various types of airplanes.

The oil pump is an engine-driven or electric-motor-driven unit. It provides the necessary pressure and flow for operating the servos and moving the control surfaces. The vacuum pump, for driving the gyros and operating the air pickoffs and air relays, is also engine driven. In many installations this pump supplies the vacuum for all other vacuum-operated instruments as well as for the automatic pilot

The sump acts as an oil reservoir for the hydraulic system. It has a sight gage for checking the oil level, and a regulator that controls the oil pressure from the pump. This regulator is adjustable to the proper pressure for the airplane in which it's installed. An oil filter keeps the oil in the hydraulic system cleaned, and is removable for cleaning—without the necessity of disconnecting any pipings or fittings.

The servo units consist of three hydraulic cylinders that actually do physical labor in the automatic pilot system. These cylinders are in a single metal casting and are generally the type in which the pistons are centered when the airplane control surfaces are centered. The pistons are connected directly to the main control cables of the airplane. As you've seen, oil pressure is applied to one or the other side of a piston to cause motion of the control in the direction desired.

The correct type of oil must be used in a servo. Various servos use different types of oil. Check to be sure you're using the oil specified for the servo you're servicing.

An air filter in the intake line to the automatic pilot control units maintains a flow of clean air through the gyros and the air pickoff system. The filter element can be removed for cleaning or be replaced without disconnecting any pipe or fittings.

Type S-3 Installation Pointers

The mounting unit of the type S-3 automatic pilot is installed behind the instrument panel in such a position that the control knobs on the face of the unit are within reach of one of the pilots. It should be placed as high as possible, and in such a position that the faces of the directional gyro and the bank-and-climb gyro control units are level in normal flight position, and are flush with the instrument panel. It's quite important that the control units can be seen easily, that they are shock-mounted, and that the various control knobs can be reached without too much trouble. Access must be provided to the rear of the mounting unit to permit adjustment of the balanced oil valves and removal for cleaning.

The servo speed control valves are in a unit arranged so that it can be attached to the lower center of the unit, .flush with the instrument panel face. However, if such a location is not desirable or possible, the speed control valves can be installed in some other position within reach of the pilot.

The speed control valve is installed in the return line from the balanced oil valve to the sump or oil reservoir. In most types of airplanes used by the Navy, the location of automatic pilot servo units has been taken into consideration in the original design. Usually the airplane control cable installation is such that a cable from each of the three controls will pass through the location of the servo unit mounting, spaced the same distance apart as the servo piston rods. In any case, the brackets and structure to which the servo unit is attached must be strong enough to withstand the design load of any two cables applied at the same time. It is best if the

servo unit is mounted in a horizontal position. In this position there is much less possibility of air being trapped in the cylinders.

The oil sump should be placed below the level of the mounting unit to permit gravity drainage. When' this isn't possible, a drain trap must be used to insure return of drainage oil to the hydraulic system. But under no circumstances should the sump be placed more than 5 feet above the drain trap. It's also best if the sump is located on a level with, or above, the oil pump to permit proper priming—but this is not important enough to take precedence over the primary consideration which is to have gravity drainage from the balanced oil valves to the sump.

The sump should be located so that the sight gage can be seen readily, and so that the tank is in a position that allows for easy filling. If space is limited, you can remove the oil pressure regulator from the sump and install it in the pressure line between the pump and the filter. A vent line should be installed at one of the outlets on the top of the sump. If it's likely that the airplane will be used for violent maneuvers or acrobatics, the vent line should be carried down as far as the bottom of the sump on either the forward or aft side.

The vacuum relief valve should be mounted as close to the mounting unit as possible, and in a place that can be reached easily for adjustment and cleaning of the intake filter screens. It can be placed on either end of the mounting unit if you are sure that the air intake faces downward. This allows access for cleaning the screen.

The oil filter is located in the main oil pressure line between the pressure regulator and the pressure manifold on the mounting unit. This unit should be supported rigidly and installed in a vertical position so that the oil connections are at the top and can be easily removed for cleaning. The inner element is covered by a fabric boot which removes injurious matter from the oil after a new installation. This boot should be replaced after the first 2 hours of use and every 5 hours thereafter until it no longer seems to be picking up any dirt. Then you can remove the boot permanently.

The air filter unit should be placed as close as possible to the mounting unit and must be rigidly supported. It is' to be horizontally mounted with the inlet down. The filter itself must be accessible for removal, and clearance must be left to prevent obstruction at the inlet end. Be sure the plug is removed from the inlet end at the time of installation, and that the filter is mounted with the inlet port downward so that collected dirt will drop off.

The main cables, which are attached to the ends of the servo unit piston rods, should have their pulleys and guides located so that no side loads will be exerted on the piston rods. Misalignment causes leakage, uneven wear on the cylinder walls, and friction in the control system. Stops should be provided in the airplane control system and set so that the servo unit is not used as its own stop at the end of its stroke.

The followup cables are attached to the piston rods, and must be installed so that they cannot become twisted with the main cables. Follow-up cables should be as short as possible and follow the most direct route from the servo to the mounting unit. Disconnect links are provided at the servo end of all followup cables. These links provide a means for disconnecting the followup cable in case it be-

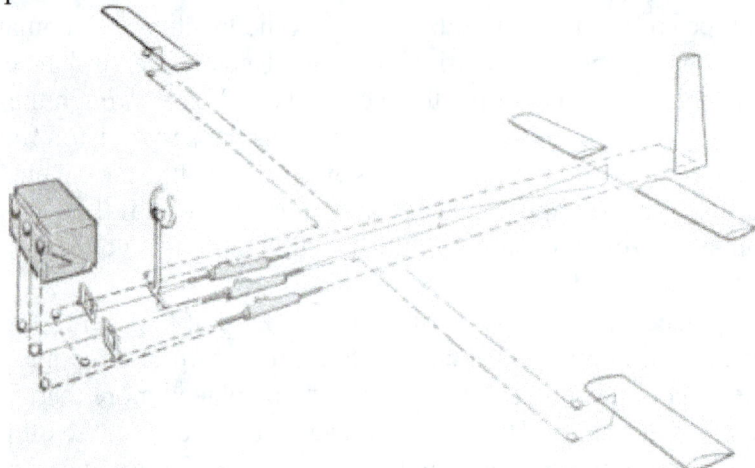

Figure 90.—Schematic arrangement of cable connections with servo unit in main control system.

conies fouled. By the application of a slight additional force on the main controls, the disconnect links, which are of the snap-type, will let go and free the control. .

Preflight Checks and Adjustments

Very likely it will be your job to conduct instrument ground tests at frequent intervals, and you should be well informed on the things that have to be done to ready the automatic pilot for service in the air. Here's a check list that will be usefully The oil level in the sump should be "three-quarters fall." If necessary, fill to the three-quarter mark with approved aircraft hydraulic oil of the type specified for servo use.

2. See that the bank-and-climb gyro control unit is uncaged by turning the caging knob counterclockwise as far as it will go.

3. Also see that the directional gyro control unit is uncaged by pulling the caging knob straight out.

4. After the engines have been started, turn the rudder knob on the directional gyro unit to aline the upper card with the lower card.

6. Turn the aileron knob until its followup index matches the zero point on the banking scale at the top of the bank-and-climb gyro dial.

6. Turn the elevator knob until its follow-index matches the elevator alinement index at the side of the bank-and-climb gyro dial. Do not aline the followup index with the horizon bar.

7. Check the vacuum. It should be 4.8 inches of mercury as indicated on the gage. At maximum r.p.m., the vacuum should not be over 5 inches. If necessary, adjust the vacuum regulating valve to obtain the proper amount. If the model of automatic pilot in question . does not have a vacuum regulating valve, adjust the vacuum relief valve at the engine-driven vacuum pump.

8. After making sure that any outside control battens are off the control surfaces, see that the emergency oil pres-

sure valve is turned on. Engage the automatic pilot by moving the engaging lever slowly all the way on. 9. On models which have servo units with hydraulically operated bypass valves instead of mechanically operated bypass valves, see that the bleed valve ie turned to normal. Then engage the automatic pilot by turning the main on-off valve to on.

10. With the engines running at 1000 r.p.m., check the oil pressure. It should be within 10 p.s.i. of that recommended by the manufacturer. This information can usually be obtained from the manufacturer's handbook for the particular type of airplane. If necessary, adjust the automatic pilot oil pressure regulator to obtain the required pressure.

11. With the engine running and the automatic pilot engaged, check for error in the hydraulic system. Check the directions in which the controls are applied to see that they aren't connected in reverse, and make certain that each control has the correct followup. The controls should not be resilient (springy) when a moderate amount of pressure is applied to them, but should feel as if they were locked. The overpower must be set high enough to apply necessary control. Do not confuse stretching of cables with air in the hydraulic system. If you're in doubt, note whether there is any movement of the followup indices of the control units when pressure is applied to the controls. Stretching of the cables will not cause these indices to move.

To remove air from the hydraulic system, set the engines at approximately 1000 r.p.m. Turn the automatic pilot oil-pressure valve on and the automatic pilot master control lever off. Center the controls and aline the followup indices. Move the controls from hard-over to hard-over, first separately and then all together. Hold each control in each of its hard-over positions for 20 or 30 seconds. Do this two or three times. In this way air in the hydraulic system is pushed by the oil flow until it reaches the sump or oil reservoir where it can escape.

On installations which have hydraulically operated bypass valves in the servo units, it is necessary to close the valve in the bypass operating line to remove air from the hydraulic system. Be sure to resafety the valve in its open position when air has been removed.

12. With the automatic pilot engaged, test its operation by rotating the rudder knob, aileron knob and elevator knob in both directions.

13. Check to be sure that the automatic pilot can be overpowered without having to apply excessive force to the controls. Try all controls each way. Then disengage the automatic pilot.

14. Reset the speed control valves to their normal operating setting. Check once more on the oil level in the sump. It should be three-quarters full, so replenish it if necessary.

ELECTRIC AUTOMATIC PILOTS P-1 Automatic Pilot
•

The P-1 automatic pilot is a system of electric automatic controls which holds the aircraft on any selected heading, brings it back without overmaneuvering when momentary displacements occur, and simultaneously keeps the ship stabilized in pitch and bank. While under automatic control the aircraft can be made to climb, dive, and execute perfectly banked turns.

This automatic pilot basically is built upon the principle of connecting an autosyn to the mechanism in the flight instruments which indicates direction, rate of turn, bank, and pitch, so that electrical impulses, or signals, will flow whenever the instrument shows a change in attitude or heading.

These electrical impulses or signals control the operation of servo units which convert electrical energy into mechanical motion. By connecting the servos to the rudder, aileron, and elevator controls, the controls can be made to move whenever signals flow in the system, thus providing a means of correcting displacements of the aircraft almost instantaneously.

180

The simplicity of control provided by the turn and pitch control permits a variety of maneuvers to be performed with minimum control knob manipulation. Normally the automatic pilot is engaged and disengaged electrically by a clutch switch, but it can also be disconnected by mechanical means in an emergency.

As long as the system is turned on, the automatic pilot is synchronized with the controls of the aircraft and functions even when disengaged. Therefore, it is always ready to take over smoothly and maintain the heading and attitude of the aircraft at the moment of transfer from manual to automatic control.

The main components of the P-l automatic pilot are a gyro flux gate transmitter and amplifier, a master direction-indicator, a bank and turn control unit, a gyro horizon control, a turn and pitch control, a servo amplifier, servos, an amplifier adapter, a power junction box, a controller selector switch, a clutch switch, and a caging relay.

These units of the automatic pilot system can be individually replaced. However, whenever a master direction-indicator is removed for overhaul and a replacement unit installed, the new indicator must be adjusted for compensation. If the compensation curve for the aircraft is available, the new indicator can be adjusted from this date; but, if any other work has been done which might have affected the deviation, the aircraft should be reswung.

Although replacement of the transmitter usually does not require recompensation of the system, it is strongly recommended that compensation be checked. Also, if the amplifier adapter is replaced, the potentiometer settings and reversing switches on the new unit must be carefully checked, and, if necessary reset, to make sure they are correct for that particular aircraft.

In the human body, signals to move us from place to place are originated in the brain as it references outside conditions. These signals are transmitted through the nerves to the muscles, and the body does its required movement by muscle power.

Similarly, as shown in figure 91, most automatic pilots have their component parts divided into three major groups, sensors, amplifiers, and servos. The P-l is no exception to this rule. The sensors consist of the units which originate the signals as they are acted upon by outside references. In the P-l the sensors are the stock-controller, the compass adapter, the rafce and rate gyro control, and the vertical gyro

VERTICAL GYRO

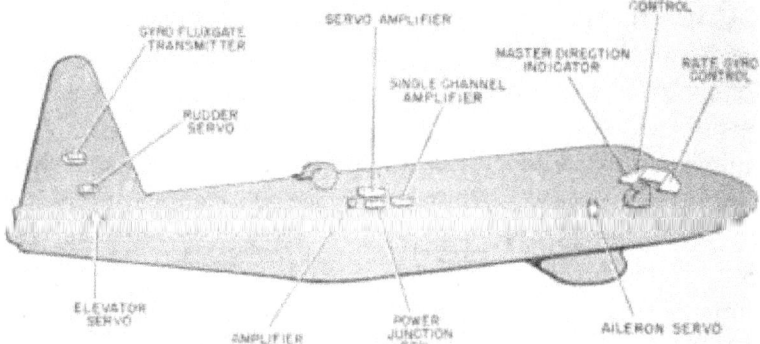

Figure 91.—Block-Pictorial schematic of the P-l automatic pilot.

control. There are two amplifiers: the servo amplifier and the compass amplifier. The automatic pilot has three servos which are the muscles of the system for putting into effect the signals from the sensors.

The Sensors

The remote compass transmitter is the directional reference unit for the entire P-l system. The compass element called the flux gate and the vertical seeking gyro, by means of which the gyro is stabilized, are housed in this unit. Upon installation the transmitter is placed in a location where it will be removed as far as possible from magnetic interference.

The remote compass transmitter consists essentially of the flux gate assembly which is stabilized by an electrically driven gyro with a ball-type erection system assembly. A caging mechanism is provided to bring the gyro assembly to a position vertical to the transmitter housing. The transmitter mechanism is enclosed in an airtight aluminum housing.

The gyro motor is of the three-phase induction type. The gyro rotor turns about the gyro stator within the gyro housing. The gyro housing is shielded from the flux gate assembly. Power at 26 volts, 400 cycles, 3-phase is supplied to the gyro stator windings to turn the rotor at an operating speed of 20,000 r.p.m.

The complete gyro assembly is mounted in a gimbal frame. The gimbal frame is mounted in a stationary frame. The assembly is designed to allow the gyro 100° of freedom of swing in bank and an average of about 70° in climb or dive as illustrated in figure 92.

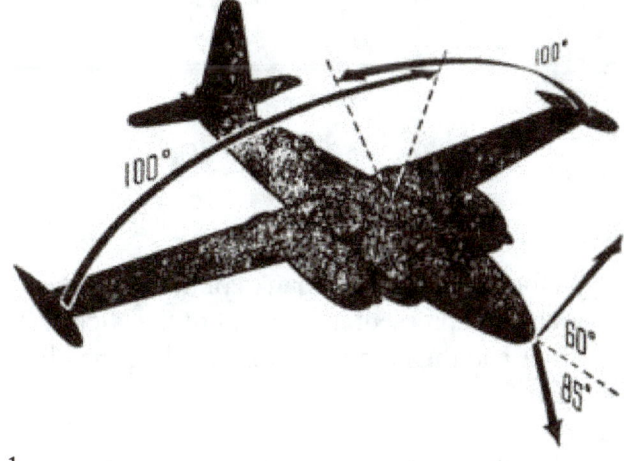

1

Figure 92.—Operating limits of P-l automatic pilot.

The vertical gyro control (gyro horizon control) furnishes the reference for the automatic pilot in attitude. It consists essentially of an electrically driven gyro motor, a bank autosyn, and a pitch autosyn. These autosyns provide corrective signals to the autopilot system. In addition, the gyro horizon control furnishes a ready visual reference to the pilot of the attitude of the aircraft.

The gyro motor of the vertical gyro control is mounted within gimbals in such a way as to allow for 100° freedom in left or right bank, 85° in dive and 60° in climb. These limits are illustrated in figure 92.

The turn and pitch controller enables the aircraft to be put through a variety of maneuvers without disengaging the

automatic pilot system. By operating the pitch trim wheel, the ship can be made to climb 40° or dive 40°; by operating the bank trim knob, it can be banked 10° in either direction. Operation of the turn control knob automatically produces a turn which, for the rate selected, is correct in both bank and pitch.

The stick-type controller furnishes the pilot with a mechanism by which he can fly the ship through his desired maneuvers without disengaging the automatic pilot, and by which he can save his strength by having the automatic pilot furnish the power by which he accomplishes these maneuvers. On the top of the stick is a detent button which must be depressed to use the stick to control the aircraft. When the detent is depressed the stick may be moved through 300° in azimuth and up to a maximum angle of approximately 30° from the vertical. Located on the top of the controller box, figure 93, are the bank and pitch trim wheels along with the plug which covers the airspeed adjustment shaft.

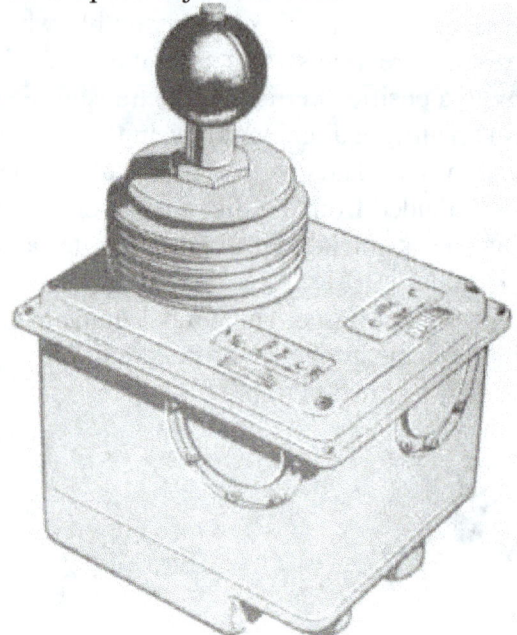

Figure 93.—Stick type controller. 184

The controller contains three autosyns for pitch, rate, and bank connected to the stick by means of a mechanical linkage. When the detent is depressed, movement of the stick is transmitted through the autosyns to the amplifier and then to the servos enabling the pilot to place the aircraft in whatever maneuver he desires.

The Amplifiers

The multiple channel amplifier embodies three separate but identical channels; one each for rudder, aileron, and elevator signals. The output of each channel furnishes power to the variable phase of the low inertia motor in the rudder, aileron, or elevator servo.

Each channel consists primarily of a 6SL7GT twin triode signal amplifier tube, the output of which is resistance-capacitance coupled to a pair of 6V6GT tubes used as phase discrimination grid-controlled rectifiers. The output of each 6V6GT is used to control a separate magnetic amplifier, the output of which is used to excite the variable phase of the low inertia motor in one of the servos.

The Servos

Since the three servos in the P-1 automatic pilot are the same in construction, a description of one will be sufficient. The actual work of driving the servo mechanism is done by the low inertia motor.

A damping motor is necessary to prevent hunting of the servo. Damping is effected by virtue of the fact that, with one phase of the damping motor under steady excitation, a voltage will be induced in the windings of the second phase, proportional to the speed of the motor. The voltage induced in the second phase windings of the damping motor is in phase to the autosyn followup signal when plane is going off course, and in phase opposition to the autosyn followup signal when plane is returning to course or attitude.

The output torque of the low inertia motor is transferred to the servo power shaft through the medium of a solenoid operated clutch. When the solenoid is energized, the clutch is held in engagement. Deenergizing the solenoid releases the clutch which is then disengaged by means of a coil spring. The clutches in the three servos as well as the one in the master direction-indicator are operated by the clutch switch.

The rotor of the followup autosyn is connected to the fol-lowup gear train, which in turn is driven by the servo mechanism, on the low inertia motor side of the clutch. As a result, the automatic pilot system is always synchronized with the ship's controls, even when the clutches are disengaged. The automatic pilot, therefore, takes over smoothly when the clutches are again engaged.

The servo disconnect acts as a mechanical coupling between the servo power shaft and the drums to which the

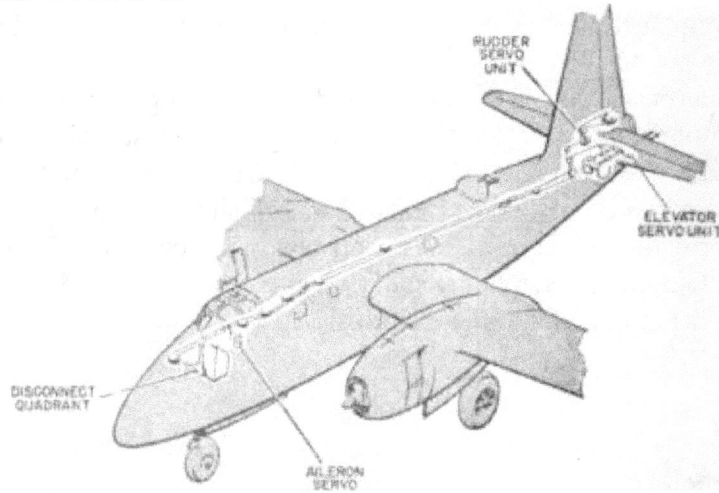

UNIT

Figure 94.—Typical layout—servo disconnect.

control cables of the craft are attached. The hub of the servo disconnect fits on the spline section of the servo power shaft while the bearing carries the slave cable pulley. Pulling the disconnect cable withdraws two drive pins from the two holes in the hub, allowing the bearing and drum to rotate freely on the hub. The servo disconnect can be reengaged by releasing the two latches protruding through the cover, and moving the controls back and forth until they drop into their respective holes. See figure 94 for a typical servo disconnect layout.

PREFLIGHT CHECK

One duty of the instrument man in a squadron will be to hold a preflight check of the automatic pilot. The check is held with aircraft power on and the automatic pilot disengaged. First, check the controls, hard over to hard over, for free motion of rudder, aileron, and elevators. During the engine warmup, check the automatic pilot after first allowing two minutes to warm up. Erect the gyros by first caging, then uncaging the knob on the vertical gyro control. With the clutch switch of the automatic pilot off, turn the automatic pilot power switch on, and allow two minutes for the amplifiers to warmup. Center the controls, trim the aircraft in the desired attitude of flight, and engage the automatic pilot by pressing the clutch switch to the on position.

The system should freely engage and disengage the control surfaces of the aircraft; and the control surfaces should move to produce climb, dive, bank, and turns in proportion to the degree of movement set into the turn and pitch controller adjustments. In each case, the direction of control caused by the movement of the controller adjustments should be checked carefully.

G-3 ELECTRIC AUTOMATIC PILOT

Automatic control of aircraft in a preset attitude and course, and also power control in maneuvering of the plane is possible through the use of the G-3 automatic pilot.

The G-3 automatic pilot consists essentially of seven electrical units connected by eight wiring harnesses. These units, which may be divided according to their functions into four types as shown in table 1, on page 188, consist of the vertical gyro control, the stick-type controller, the rate and pendulum control, the G-2 compass system, the control amplifier, the servo drives, and the power adapter.

When the desired attitude and course have been preset on the automatic pilot, deviations from that attitude or course will produce deviation signals. These, when they are amplified to a strong enough magnitude and are of the proper

Table 1.—Classification of G-3 automatic pilot components

classification, return the airplane to the desired course and attitude by means of electric servo drives.

At this point we find the purpose of the G-2 compass and the vertical gyro control unit in the automatic pilot. Course and attitude deviation signals originating in the master direction indicator and in the vertical gyro are transmitted to the proper servo drive by the control amplifier. Oscillations in yaw are damped out and the rudder is coordinated in turns by the rate and pendulum control unit as it provides deviation signals. Constant barometric altitude flight is another feature possible with the G-3 automatic pilot as the deviations sensed by the barometric altitude controller, figure 95, are incorporated into the overall signals.

PHASING CONDENSER
SOLENOID

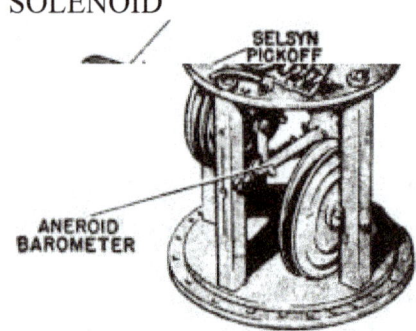

Figure 95.——Altitude controller.

188

When for any reason the pilot wishes to maneuver the aircraft with the automatic pilot, he can originate his signals for the desired maneuvers by means of the stick-type controller, figure 96, in practically the same manner as with the regular control column. These external signals are introduced into the automatic pilot causing the servo drives to move the control surfaces, and directing the airplane to a new attitude, a new course, or both as desired.

ALTITUDE

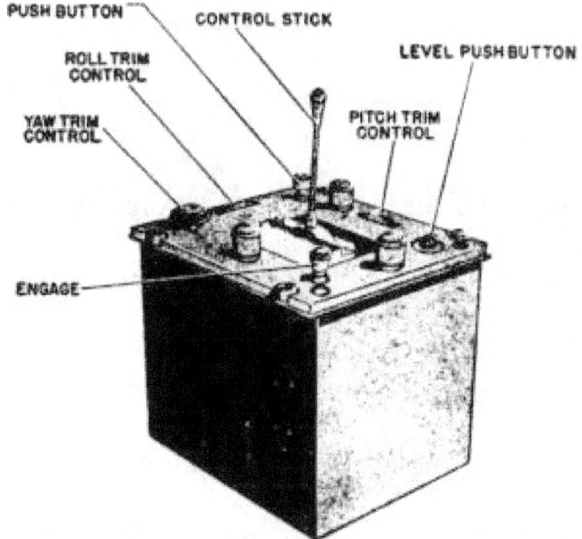

Figure 96.—Stick-type controller.

Continuous synchronization of the G-3 automatic pilot permits it to be engaged at any time the pilot desires without the danger of a sudden movement of the aircraft. Gyro drift is also prevented by the stabilization of the G-2 compass system to correspond to the magnetic heading of the plane at all times. The necessity of caging and uncaging of the gyros before or after maneuvers is eliminated by the continuous synchronization and stabilization of the gyro components.

The G-3 automatic pilot has exceptionally good operating limits. These limits are 70° in roll, climb, and dive and 360° in course or yaw as illustrated in figure 97. By means of the stick-type controller, the aircraft can be maneuvered through these full operating ranges. Coordinated turns can also be made without any appreciable change in altitude.

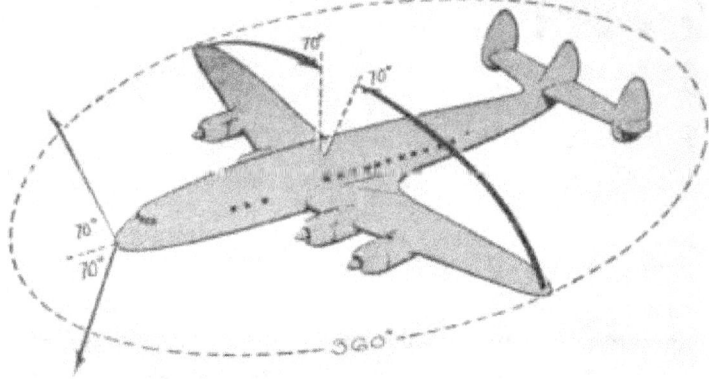

Figure 97.—G-3 automatic pilot operating limits.

For added safety, if the pilot should need to take over the controls of the aircraft and has no time to shut off the automatic pilot, he can overpower it by operating his manual controls with a greater than normal force.

VERTICAL GYRO CONTROL UNIT

The vertical gyro control unit is essentially a free gyroscope. In other words, the gyro rotor is mounted in a gimbal so that it can move in any direction. It is mounted in the aircraft so that its rotor spin axis is in a vertical position and the pivot axis of both the inner and outer gimbal rings lie in the horizontal plane when the aircraft is in a level flight attitude.

When the aircraft moves around either of its horizontal axes, it changes its position with respect to the gyro spin axis. This change in position is detected and measured by means of selsyn pickoffs attached to the gyro so that they generate signals proportional to the relative movement of the gyro gimbal rings with respect to their mountings. These signals are proportional to the amount the aircraft deviates from its original attitude and may be used in a servo-mechanism system to bring the aircraft back to its original attitude.

nr.

Two three-terminal selsyn piekoffs are used in the vertical gyro control unit; one to detect deviations about the longitudinal axis of the aircraft, the other about the lateral axis. The gyro is kept erect by the gyro erection system which consists essentially of two pendulums and two torque motors, mounted on the two horizontal axes. In normal unacceler-ated, level flight, any inequality between the outputs of the pendulum and gyro piekoffs on either axis indicates that the gyro axis is off the vertical by an amount proportional to the error signal. These error signals are fed to an erection am-

plifier which, in turn, energizes the proper torque motors to return the gyro spin axis to the vertical.

A miniature stick for maneuvering the aircraft and knobs for trimming the aircraft when the automatic pilot is engaged, are located in the stick-type controller of the G-3 automatic pilot. This controller is the pilot's control station for the G-3 automatic pilot. It also contains the switches that engage the automatic pilot, the return-to-level-flight system, and the altitude control system.

The stick mounted on the controller is geared to the roll

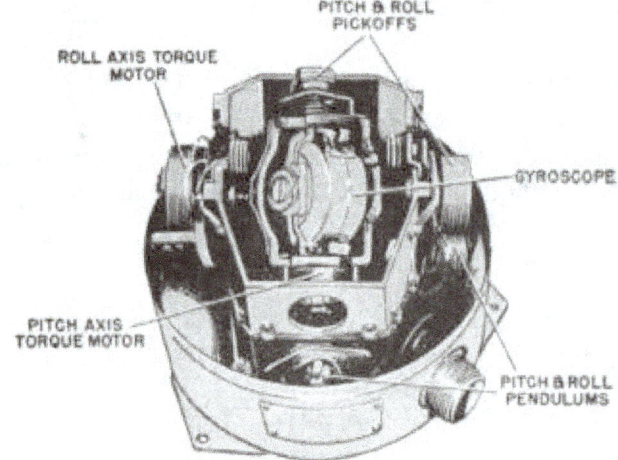

Figure 98.—Vertical gyro control unit.

STICK-TYPE CONTROLLER

and pitch maneuvering selsyn pickoffs. Motion of the stick about the fore and aft axis causes the aircraft to maneuver in roll. Motion about the lateral axis causes the aircraft to pitch. Basically it is a position-type controller; that is, the angles attained in a maneuver are proportional to the stick displacement from neutral.

The turn compensation selsyn pickoff is also actuated when the maneuvering stick is moved in the roll direction. This pickoff generates a signal proportional to the pitch attitude change required for the necessary increase in wing lift during a banked turn. The signal is fed through the amplifier to the elevator servo to increase the airplane's angle of attack so that the wing lift is increased and no altitude is lost.

Rate and Pendulum Control Unit

The rate and pendulum control unit consists of a rate gyro and a pendulum control unit mounted in one container. The rate gyro operates primarily as a yaw damper as it damps

GYRO ASSEMBLY

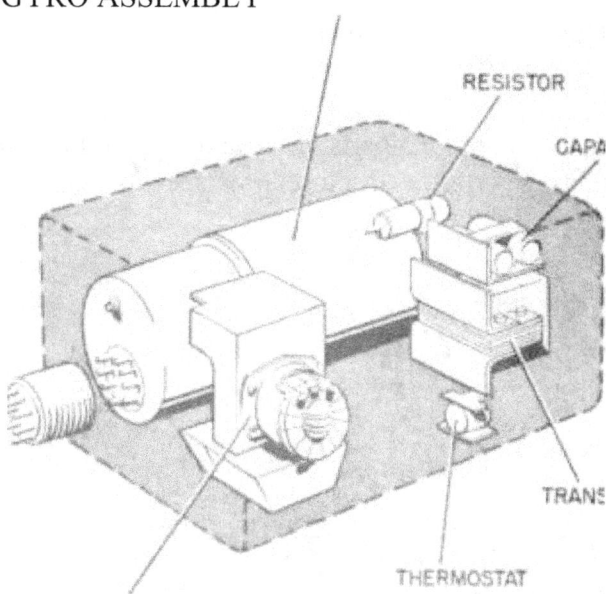

RESISTOR

CAPA

TRANS

THERMOSTAT

CAPACITORS
TRANSFORMER
PENDULUM ASSEMBLY
THERMOSTAT
and pendulum control unit* 193
out airplane oscillations about the vertical axis. The pendulum coordinates the rudder in maneuvers.

When the airplane deviates in yaw, the rate gyro pickoff generates a signal which is proportional to the rate of deviation of the gyro about the yaw axis. This signal is fed to the rudder servo amplifier. When this signal is fed through an amplifier to the rudder servo, the rudder servo drives the rudder in the proper direction to decrease the rate of deviation. In this way the rate gyro furnishes signals that are used to damp out oscillations about the yaw axis.

The pendulum control unit operates in the system at all times to coordinate the rudder with the other controls. Consisting essentially of a pendulum subject to two forces, gravity and centrifugal force, as the plane rolls about its longitudinal axis, the pendulum originates a signal which is transmitted to the rudder servo to turn the airplane in the proper direction and at the proper rate.

Servo Drives

There are three identical servo drives in the G-3 automatic pilot: one for the rudder, one for the elevator, and the third for the ailerons. Each consists of a split-series d-c motor, gear reduction, followup selsyn pickoff, tachometer, solenoid-operated clutch, and output pulley. When the autopilot is engaged, the pickoff and the output pulley are both geared to the motor shaft. The solenoid-operated clutch mechanically disconnects the pulley from the gear train and motor when the automatic pilot is disengaged. The pulley is coupled to the control surface of the airplane at all times.

A distinct advantage over hydraulic automatic pilots can be seen in this portion of the G-3 automatic pilot in the positioning of the servos near the control surface which they actuate. This is graphically illustrated in figure 100.

Power Adapter

The power adapter is the electrical connecting link between the G-2 compass system, which was discussed in an earlier

chapter, and the power supply to the automatic pilot. It consists of two transformers mounted in a hermetically sealed can.

Line Maintenance

It sometimes becomes necessary to remove certain parts of the automatic pilot in the maintenance of the gear. With few exceptions, any defects disclosed in inspection are remedied by the substitution of a new or overhauled unit for the defective one.

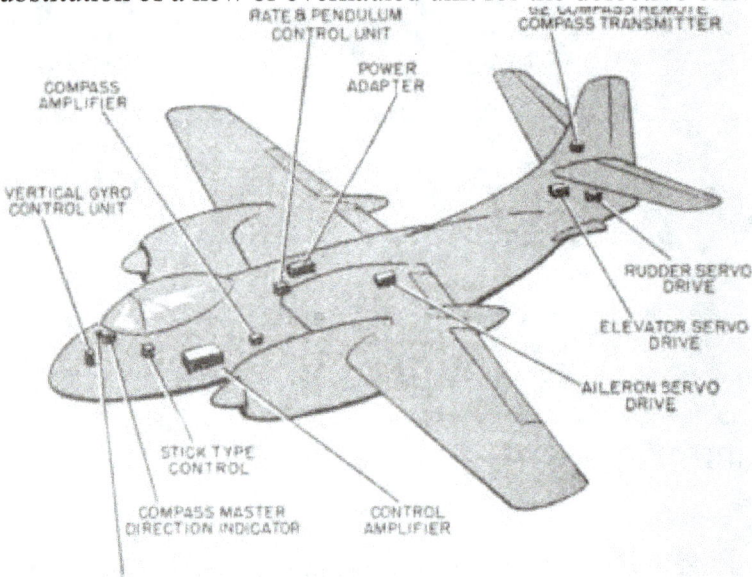

EMERGENCY DISCONNECTION

Figure 100.—Location of the various units of the G-3 automatic pilot.

Whenever a unit must be removed from its location in the airplane to make it accessible for inspection, be sure that all inspection is completed before the unit is reinstalled. In particular, be exceedingly careful in handling gyros since they are among the most delicate instruments made. When a gyro is removed from its mountings, set it on a cushion of felt or other soft material. Never remove the covers from gyros. It is also imperative to remember that dirt, the great-

est single enemy of any moving mechanism, should be kept out of the units that comprise the automatic pilot.

QUIZ

1. The S-3 automatic pilot combines with pneumatic and hydraulic mechanisms to work the controls in coordination.

2. The S-3 bank and climb gyro spins with its axis

3. The S-3 followup goes into action while the airplane .

4. What devices allow the human pilot to overcome the S-3 automatic pilot?

5. To use the stick type controller on the P-l automatic pilot, the must be depressed,

a. Clutch switch.

b. Detent.

c. Pickle switch.

d. Caging knob.

6. The P-l automatic pilot should be warmed up for minutes before engaging.

7. In ground checking before the automatic pilot is engaged, the controls should be

checked to for freedom of movement

8. When the pilot desires to maneuver the aircraft with the Q-3 automatic pilot, he can originate his signals by means of the

9. Operating limits of the G-3 automatic pilot are (a) degrees roll, (b) degrees climb, (c) degrees dive, (d) degrees in course.

10. The G-3 gyro erection system consists essentially of and

11. govern the speed with which the automatic pilot operates the controls.

12. The device for setting the S-3 automatic pilot for any desired heading is located in the unit.

13. To compensate for variation in load conditions for nose up or nose down in the S-3 automatic pilot, the miniature airplane

14. Speed control valves of the S-3 automatic pilot are installed in the

15. Normally the P-l automatic pilot is engaged and disengaged electrically by a

\'7d6. Whenever a P-l automatic pilot master direction-indicator is removed and a replacement unit installed, the new indicator must be adjusted for .

17. The P-l automatic pilot has two amplifiers: and

18. The output torque of the low-inertia servo motor of the P-l automatic pilot is transferred to the servo power shaft through a clutch.

19. Gyro drift of the G-3 automatic pilot is prevented by the stabiliza- | tion of the G-2 compass system to at all times.

»

20. Basically the stick-type controller of the G-3 automatic pilot is | a controller.

21. The S-3 automatic pilot provides visual indications in all but
a. Yaw. I
b. Altitude.
c. Pitch.
d. Roll.

22. The valve is used to turn the S-3 automatic pilot on and off.

23. The position of the follow-up card of the S-3 automatic pilot is controlled by the above the

24. The S-3 servo units consist of three that do the labor of the automatic pilot system.
i

25. The P-l automatic pilot is built basically on the principle of connecting a(n) to the flight instrument mechanisms.

26. The P-l automatic pilot is with the controls of the aircraft as long as the system is turned on.

27. Most automatic pilots component parts are divided into three major groups; is not one of them.
a. Incisors.
b. Sensors.

c. Amplifiers.

d. Servos.

28. The is the directional reference for the entire P-l
automatic pilot system.

2& The barometric altitude controller of the G-3 automatic pilot makes possible.

30. The pendulum of the rate-and-pendulum control unit

a. Acts as a yaw damper.

b. Coordinates the ailerons in maneuvers.

c. Coordinates the rudder in maneuvers.

CHAPTER

NAVIGATION INSTRUMENTS DRIFT SIGHTS

Dead reckoning navigation, as you probably know, is determining where your airplane is by taking into account where you started your flight, and the distance and the direction in which you have flown. Sounds easy ? Actually, it's no snap job when you consider all the factors that can affect your calculations on distance and direction. Precision, know-how, and good instruments are essential.

For example, whenever the wind through which you fly does not blow exactly parallel to the course you have set, your airplane will drift to the left or the right. So, you have to fly a compass course that has been corrected for wind speed and direction. In other words, the nose of your airplane won't point toward your destination a large part of the time, and you have to figure out the correct angle to allow for drift.

Proper allowance for the effect of wind on the heading of your airplane is determined in many instances by the use of an instrument called a drift sight, an example of which may be seen in figure 101.

The principal parts of the Navy's Mark 2B pelorus drift sight are the sighting tube, support tube, and pointer, making up one assembly; and a base plate assembly—usually two or more for each plane.

The sighting tube assembly fits into the base plate and is secured by a spring-pin. This pin holds the sighting tube assembly firmly in place, but allows it to be turned. On top of the base plate assembly is a pelorus ring—marked in

degrees from 1° to 360°. This ring also rotates independently, but can be locked in any position by a screw stud. This provides a means of setting the ring to indicate true, relative, or compass bearings.

V

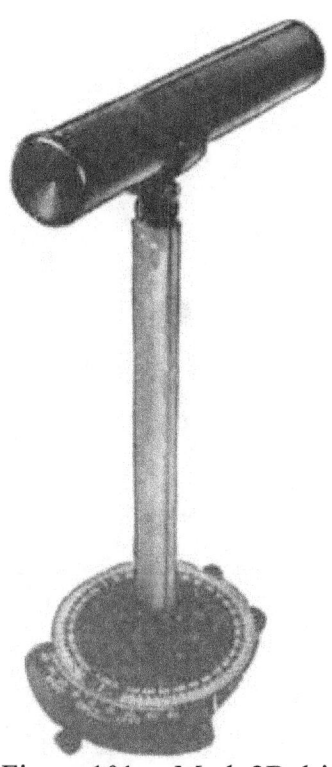

Figure 101.—Mark 2B drift sight.

The pointer, which turns with the sighting tube assembly and is located just above the pelorus ring of the base plate, has two reference markers. One is a taut wire for reading the bearing on the pelorus ring. The other is a mark for taking readings on the drift scale. If you look at the illustration again you'll see that the drift scale is marked in single degrees from 0° to 40° left and right, and that one side of the scale is marked right, the other left (or, on older instruments, plus and minus).

Open vane-type sights are mounted topside on the sighting tube so you can locate objects roughly before using the small

aperture in the sighting tube to find them accurately. The supporting tube is adjustable in height, and locks in place in the base collar by means of a screw stud. Very similar to the Mark 2B in all but the sighting head is the Mark 2C, which includes several improvements and additional accessories. A Mark 2C drift sight is shown in figure 102.

Figure 102.—Mark 2C drift sight.

When flying over land, you measure drift by observing the angle at which the airplane is flying relative to various landmarks below it. If you're over water, you create your own "landmark" by dropping a smoke light, and take sights on it by means of the drift-sight instrument. It's necessary, of course, that the speed and heading of the airplane be held constant while the sight is being taken.

If the wind is causing the airplane to drift—and it usually is unless you're flying with a direct headwind or tailwind—

the movement of the plane will appear to be at an angle. In other words, the direction in which the airplane is pointing will be somewhat different from the direction in which it is actually traveling. The angle between the heading and the course being followed (or track) is the drift angle. A drift sight enables you to measure this angle, and thus provides an important piece of information for accurate navigation.

The Mark 6 drift sight is another—and somewhat different—type of drift sight used in Navy airplanes. It performs the same chores as the types just described. The optical equipment in the Mark 6 is housed in a vertical tube which is an extension of the supporting post. The upper and lower halves of the tube together form a sort of telescope which rotates in a fixed mounting sleeve, and drift scales are mounted on the mounting sleeve. The reference pointer is attached to the top of the lower tube section and is fitted with a stylus-type marking device which not only provides you with a drift angle reading but marks the reading down on paper. Reference to figure 103 will show you what the Mark 6 looks like.

In installing a drift sight, make certain that it is placed in such a location that the instrument can be rotated horizontally through 360° without hitting any obstructions. You also have to be careful that no parts of the airplane will interfere with the user's line of sight. The mounting should be vertical when the airplane is in its normal flight position, and be supported on shock-mount members so that vibration will not blur the navigator's vision. The drift sight is mounted so the fore-and-aft center lines of the base are parallel to the airplane's fore-and-aft axis, with the drift scale forward.

In looking through the eyepiece of a Mark 6 drift sight, youH see more than the ground or water below. The picture will appear to be crossed by a series of parallel lines called the grid lines. These lines are set exactly parallel to the fore-and-aft axis of the airplane when the drift reading is

Figure 103.—Mark 6 drift sight.

zero, but can be turned so they'll be alined with the "flow" of the earth's surface directly beneath the plane.

When the drift scale is in the correct fore-and-aft position, the mounting sleeve can be locked in place. The exposed lenses of the drift sight should be inspected for cleanliness every day. You should also check the instrument at regular intervals to see that it's mounted securely, alined properly, and that it operates freely. Always look for broken or damaged parts during instrument inspections, and see that the lighting system—if any—is in working order.

THE ASTRO-COMPASS

The astro-compass provides an aerial navigator with a means of finding the true heading of his airplane by reference to the sun, the moon, or the stars. It is an astronomical instrument, and makes no use of such familiar direction-indicating devices as magnetic needles or gyros in its operation.

Primarily, the astro-compass is an adjustable sighting assembly attached to an azimuth circle which is free to rotate against a lubber's line labeled "true course" or "true heading." The azimuth circle is mounted on a fitting which is inserted in the standard, or base. You can keep the azimuth circle level by adjusting two screws and referring to two bubble levels lying horizontal and at right angles to one another. You can readily identify the various parts of the astro-compass by referring to figure 104.

The sighting assembly at the top of the instrument is adjustable for geographical latitude, local hour angle, and declination —three terms familiar to navigators but difficult to explain briefly. It's enough, for here, to say that these factors must be known to the navigator, and that adjustments for each must be applied before astro-compass readings can-be taken.

Two flanges, firmly fastened to the azimuth circle, support an axis on which the upper parts of the instrument are mounted. One end of this axis is fitted with a worm gear hand wheel by which the astro-compass may be set for lati-

tude. A scale, reading in degrees of latitude, is marked on a disk attached to the worm wheel, and another scale is engraved on a drum attached to the worm shaft. The scales for north latitudes are colored white; the scales for south latitudes are marked in red.

Figure 104.—The astro-compass.

The axis suspends a drum with two movable end plates— one on the top and another on the bottom of the drum. Hour angle scales are engraved around the edges of the two movable plates, which turn together with reference to the stationary drum between them. The plates are turned by an internal bevel gear, operated by a knob placed at the end of the axis opposite the latitude worm wheel.

The local hour angle scale that the navigator uses in north latitudes is on the upper plate and is colored white. It is read against a white index mark on the drum, labeled

"L. H. A., N. Lat." The scale for south latitude is marked in red on the lower plate, and is read against a red index mark on the opposite side of the drum.

When the astro-compass has been set for the correct latitude, the drum carrying the hour angle scale is parallel to the plane of the celestial equator. If you haven't studied navigation, rest assured that the navigator in your airplane knows what this means—and don't you worry about it.

Attached to the upper face of the hour angle drum, and alined parallel to its 0°—180° line, is a sighting assembly. This consists of a shadow pin, a screen, and a star sight. The assembly can be titled relative to the hour angle plates, and can be set for declination from 64° north to 64° south. A pointer on the sighting assembly gives a declination reading against a scale marked on an arc that is firmly fastened to the upper hour angle plate.

The shadow pin is at the end of the sighting assembly that is above the 0° mark on the white hour angle scale, and the screen is at the end over the 180° mark. The astro-compass is alined on the sun when the shadow of the pin falls between two parallel lines on the screen. The star sight consists of a lens mounted above the screen, and a fore sight between two luminous lines above the shadow pin.

When you install an astro-compass, be sure it is mounted so that the fore-and-aft marks on the mounting base are parallel to the fore-and-aft line of the airplane, and that the instrument is fastened securely. Over a period of time the leveling screws may become loose, so they should be checked periodically. If the navigator reports that the azimuth circle turns too freely, its motion can be tightened by adjusting the center nuts on the under side of the instrument with a spin-tite socket wrench until the desired amount of friction is obtained.

THE ASTROGRAPH

The astrograph is an instrument that resembles, more or less, a magic lantern or film slide projector. It has, however, a very specialized job to do, and can't be used to show scenic views of the Grand Canyon or the Pyramids. Its purpose

is to project curved lines of shadow onto a navigation chart. These curves provide the navigator with information of great value to him when he's checking the position of his airplane by means of heavenly bodies, and the use of his chart as a projection screen saves him a great deal of time that otherwise would be consumed in looking up information about star positions in books. You see an astrograph, but no chart, in figure 105.

UUMJNATON CONTROL

on off swrrcM

Figure 105.—The astrograph.

Like the astro-compass and certain other celestial navigation instruments, the astrograph must be adjusted properly with relation to the geographical latitude and the local time before use. In addition, it must be set so that the size of the image projected by it is in the same scale as the chart.

The two main parts of the astrograph are a mounting ring, fastened firmly above the navigator's table, and a detachable projector. The projector is supplied with a height gage, two kinds of projection lamps, seven cans of film covering different positions of latitude, and a takeup spool like those on which the film is wound.

Two different projection lamps are supplied with the astro-graph. The use of one or the other depends upon the height of the projector above the chart. When the projector can be used at a distance of 22.3 inches above the chart, for example, the shorter projection lamp is used. But when the projector is used at 16.3 inches above the chart, the long projection lamp is used. The height of the projector above the chart depends entirely on the available room in the airplane.

The projector is attached to the mounting ring by three spring clips, and you can adjust the projector for height and levelness by three screws on the ring. Two wheels on the side of the instrument enable you to wind the film to any desired east-west position. Between the wheels is a small knob for adjusting the north-south position of the curves. Close by the N-S adjusting knob is another knob which operates a rheostat (or variable electrical resistance) to control the intensity of light projected. On one make of astrograph the rheostat knob is also used as an "on-off" switch. Another type has a separate "on-off" toggle switch near the knob.

All the astrograph equipment—including the mounts, instrument, film, lamps, and necessary wrenches for adjustment—fits into a carrying case made of plywood. The case is equipped with brackets so that it may be secured to the floor of the airplane during flight by means of a one-inch strap.

When you're installing an astrograph, be sure that the mounting ring assembly is positioned so that it is parallel with the navigator's table. N The support must be firm and rigidly attached so there will be as little vibration as possible. . And it should be centered above the table so that light from the projector will have to travel the same distance to either side of the mounted chart. Above all, be positive that there are no obstructions between the projector and the table to block the beams of light.

To mount the instrument, hook the two mounting clamps on its left side over the corresponding clevis pins of the mounting ring. Using the clamps as hinges, swing the instrument into position. Be careful to see that the leveling screws seat properly upon the hole, slot, and mounting pads of the housing. Then engage the right-hand mounting clamp with the remaining clevis pin. Attach the electrical receptacle to the plugs on the rear panel of the housing. The receptacle can be suspended from the retaining plate of the mounting ring when it's not attached to the projector.

Before each flight you should clean the aperture glasses of the lamps with a soft cloth to remove smears, dust and fingerprints. Wipe the threads of the lamp holder, and check to see that the holder turns smoothly in the lamp house bushing. Inspect the facing of the springs which hold the film and film guide rollers, to see that they are clean. Test the lamps by placing them in the instrument, not forgetting to check the spare. Be sure, too, that the lamp filaments are not too far off center by watching the movement of the projected image as you rotate the lamp. Check the frame's retaining screws to see that they are tightly in place.

The spring contact in the lamp house and the fluted parts of the lamp holders must be kept free of corrosion. Wipe them occasionally with an oily cloth. Oil the spool drive shafts and mounting clamp rods with a small amount of light lubricant every now and then, but do not forget to wipe off excess oil to prevent it from accumulating dirt.

SEXTANTS

The sextant is one of the most important instruments used by navigators. It is fundamentally a kind of telescope for measuring the angular height of a celestial body—such as the sun, a planet, or a star—above the horizon. This height is not measured in feet, or miles, or inches, but in terms of angular degrees. When the navigator knows this height angle on two or more celestial bodies, he is in possession of information that enables him to plot his position on the face of the earth with remarkable accuracy.

There are a number of different types of sextants in use for various navigation needs. Experience has shown that those varieties known as "bubble sextants" are the most practical for use in airplanes. The bubble in the bubble sextant provides the navigator with an artificial horizon from which to measure his angular altitudes. An artificial horizon must be used when the natural horizon is not visible because of clouds or darkness, or is obscured by haze. In practice, you'll find that the bubble horizon comes into use a high percentage of the time.

PERISCOPIC SEXTANT

The Periscopic Sextant is designed primarily for use in high-speed pressurized aircraft. However, its accuracy is in no way impaired when used under other conditions. It completely eliminates the costly and unstable astrodome by permitting observation of celestial bodies through a peri-scopic system without encountering the dangers of errors usually associated with the astrodome.

To accommodate the periscopic sextant, a mount is affixed to the roof of the aircraft cabin. When the sextant is not in use, the mount may be sealed by a shutter which is flush with the aircraft skin. When the sextant is to be used, the sealing shutter is drawn aside and the sextant extended so that the tip of the tube is exposed iy 2 inches above the aircraft.

Figure 106.—Periscopic sextant mount.

This sextant has a gimbal support whereby the navigator may compensate for variations within 15° of normal aircraft attitude and the sextant is rot at able through 360°. An independently rotatable compass rose is engraved in increments of 2° on the azimuth ring, and it is provided with two indices, one fixed, indicating the aircraft heading, the other rotatable with the sextant, to indicate the direction of the observed body. Thus, the mounting of the sextant permits the determination of both the relative or true bearing and the altitude of a celestial body.

A mechanism, called an averager, has been designed to interpret the action of a moving ball across a rotating disk and, as a result of this movement, accurately compute an average angle for any time interval up to 2 minutes. This computer is used with the periscopic sextant. At the end of 2 minutes an automatic device causes a shutter to block the optical path, thus indicating that the observation has been concluded.

An index prism in the top of the periscopic tube is ro-tatable about a horizontal axis to permit observations at any angle from —10° to +92° in elevation as compared to an artificial horizon. The optical system of the sextant is a two-power telescope with a true field of 14.5°. This wide field facilitates the location and identification of celestial bodies. All glass-air surfaces are coated to minimize light losses caused by reflections. The eyepiece is adjustable to focus from —2 to +2 diopters (a unit of board length). Filtering glasses of eight values are provided for selective use in the optical system so that the intensity of the sun's light can be adequately reduced.

To prevent condensation when the tip of the sextant is extended into cold air, the tube is filled with dry air and sealed, and as a further precaution, a small amount of indicator silica gel, visible through the objective window, provides a warning of the presence of moisture. A sealed outer tube encases the optics tube and mechanism and so protects them from shock due to normal handling. Also it lessens the effects of changing ambient temperatures.

The periscopic sextant is located in aircraft in such a way that the lever operation of the mount is not hampered by structural braces of the aircraft. In installing, the heading index of the mount must be alined with the longitudinal axis of the aircraft. This requires use of the sextant. The tube of the sextant is inserted into the mount as far as possible and the lower ring of the mount rotated clockwise

until the lug on the tube enters the slot in the mount. A retractable plunger is pulled and the lower ring of the mount rotated counterclockwise until the sextant is free to rise vertically. The shutter is withdrawn by moving the lever of the mount to the open position. The sextant is then

Figure 107.—Periscopic sextant.

pushed into the mount until the retractable plunger knob snaps into place.

Caution should be exercised in installing the sextant to

prevent its being forced against the shutter of the mount and thus damaging the tip of the sextant tube or the shutter. Also the sextant should not be removed from the mount until the shutter is closed. When flying in turbulent air, the sextant should not be allowed to remain in the mount in a retracted position.

With the sextant installed in the mount, properly alined, and the battery cable attached to the electrical receptacle located in the bottom of the mechanism, observations may be made. In making an observation, the images of the celestial body and bubble in the bubble cell should be as nearly as possible at the center of the field as indicated by the crossed reticule lines.

The sextant, if carefully handled, should require little maintenance. Periodic inspection and lubrication of the grooved track of the shutter is necessary. The frequency of inspection is dependent upon the climatic conditions encountered by the aircraft.

If either the artificial horizon, time dial or counter should not be visible at night, the probable cause is a defective bulb, disconnected or weak power supply, or a broken lead. Steps should be taken to remedy this condition by the replacement of parts.

Should moisture form inside the objective window, or if the silica gel indicates a moisture content, gaskets have become faulty and the sextant should be returned to a repair depot. Practically all other operational troubles will require returning the instrument to overhaul for repair.

MARK 5 SEXTANT

The Mark 5 sextant is used by the Navy and is shown in figure 108. It enables you to measure the angular altitude of a celestial body by reflection of its image through a totally reflecting prism. The Mark 5 has an elaborate optical system that is similar to that originally used in a Mark 4. The Mark 5 is quite streamlined and has a different timing mechanism than the Mark 4.

The Mark 5 has a chronometric averaging device which

automatically indicates the average angle of altitude of an observed celestial body during a definite period of time. It consists of three units—a gear train rotated by the scale drum which measures the change of the angle of observation with

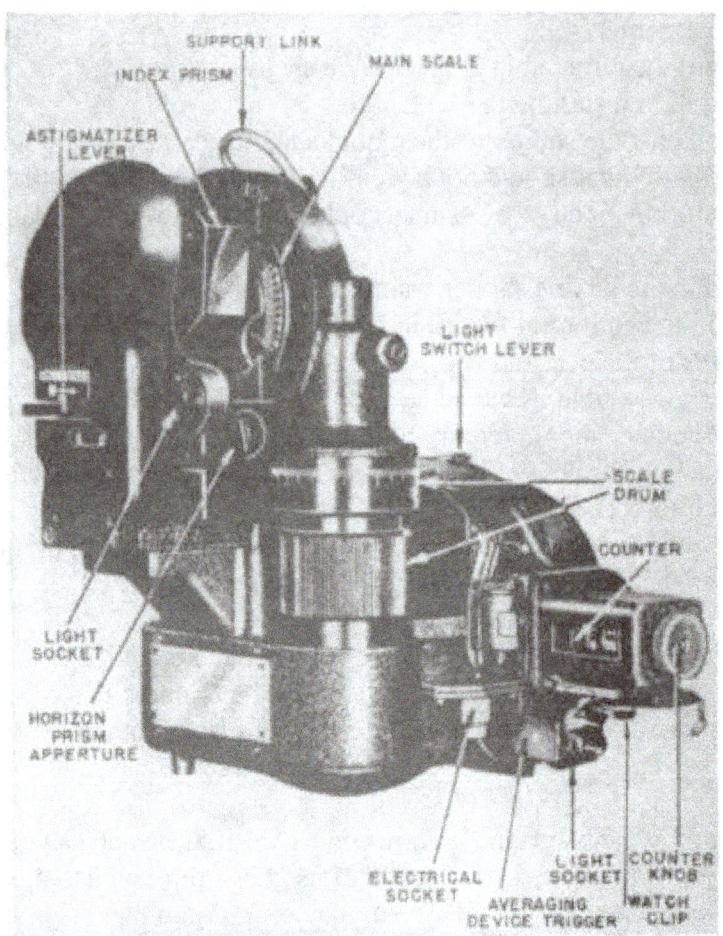

Figure 108.—Mark 5 sextant.

respect to a base-line reading set into the mechanism, a spring-driven clock mechanism for timing and driving a transfer unit; and a recording unit, operated by the transfer unit, to indicate the average of the angular changes during the time of observation. The recording unit is made up of two wheels with gradua-

tion marks for reading angles. The first wheel reads from 0° to 9°; the second wheel reads minutes of angle from 0' to 59'. The final reading is determined by the counter with the main scale as a guide. After 2 minutes running time, a shutter prevents further observation until the mechanism has been rewound with a winding key. Rewinding automatically opens the shutter for the next observation. A counter knob is turned to reset the recording unit to zero. If the navigator wishes, he can use the instrument without the averaging device by pushing the proper lever.

Before the sextant is ready for use in flight, it should be inspected to make sure the bubble can be formed properly, that the lights function, and that the controls are in working order. You should also check the running time of the averaging device clockworks. See that the main scale index line is lined up with a multiple of 10 on the main scale when the drum has been turned down to a stop. And check the indications of the averaging device against a fixed altitude setting. On both the Mark 4 and Mark 5 sextants it is important to remember that the bubble control nut must be free, with no tension on the bubble diaphragm, before the sextant is stowed away.

MEDIAN TYPE SEXTANT

Another kind of sextant sometimes used by the Navy is the median type, which is constructed so that several observations can be made on any one celestial body and the median (or average) of these several sights can be obtained from one reading of the scale. This is made possible by the use of a recording drum upon which a trigger pencil makes a penciled line for each of the various "shots" taken on the sun or star by the navigator. The middle one of these marks is then read on the scale—and it's the median of them all. By using additional marking drums, rapid observations on one or more additional stars can be made in order to get a position "fix." A median sextant is shown in figure 109.

The median type sextant also differs from the Mark 4 and Mark 5 in that it uses mirrors instead of prisms for the internal reflection of light through the component optical parts of the eyepiece. A plane and parallel-sided glass plate (the index mirror) is actuated by the same arm mechanism which carries the vernier scale from which angle readings are obtained.

You read the vernier scale against a segment of a circle that is marked in graduations from 0° to 90°. The vernier plate is graduated in 2 divisions. The arm and vernier plate are moved along the scale by a gear train and a hand-operated drum.

Figure 109.—Median type sextant.

A lighting system provides illumination for the scale and vernier. A separate, removable lamp assembly lights the artificial horizon bubble chamber for night observations, and daytime illumination is obtained through a glass window from the top of the sextant. Two sunshades are also part of the lighting system. These are mounted in a swinging bracket so they can be placed in position when you want them.

If you're in charge of maintenance on a median sextant, replace defective illuminating bulbs and batteries when you find them. Always remove the batteries from the sextant when a flying mission has been completed. Check to see that the bubble housing assembly works properly, and replace it if defective. Clean the index mirror if it's dirty or fingermarked, removing dust with a camels-hair brush. And handle the mirror with care. The objective lens also rates the same cleaning and care.

QUIZ

1. A drift sight is used to determine the proper allowance for the effect of

2. When flying over land, drift is measured by observing the angle at which the airplane is flying

3. The angle between the and the is the drift angle.

4. The provides an aerial navigator with a means of finding the true heading of his airplane by reference to the heavenly bodies.

5. When installing an astro-compass, be sure it is mounted with the marks parallel with the fore-and-aft line of the plane.

6. The astrograph projects onto the navigation chart.

7. Before each flight you should clean the aperture glasses of the astrograph lamps with
a. Carbon-tetrachloride.
b. Carbon-disulphide.
c. Soap and water.
d. A soft cloth.

8. A is an instrument for measuring the angular height of a celestial body.

9. When a flying mission has been completed, from the median sextant.
a. Never remove the batteries.
b. Always remove the batteries.
c. Usually remove the batteries.
d. Disconnect the lamps.
274033°—54 15 217

10. What are mounted on top of the sighting tube of a drift sight, so objects can be located roughly?

11. The exposed lenses of the drift sight should be cleaned

12. The sighting assembly of the astro-compass is not adjustable for
a. Latitude.
b. Local hour angle.
c. Declination.
d. Longitude.

13. The scales for south latitude on the astro-compass are colored
a. Red.
b. White.
c. Blue.
d. Green.

14. The height of the astrograph projector above the chart
a. Is 22.3 inches.
b. Is 16.3 inches.
c. Is not more than 20 inches.
d. Depends upon the room.

15. The most practical sextants for use in aircraft are the type known as sextants.

16. The ,, on the Mark 5 sextant elongates the image of the celestial body to a band of light.

17. A landmark for drift sights over water may be made by dropping

18. In installing a drift sight, make sure it will rotate degrees.

19. The astro-compass is a(n) instrument.

a. Magnetic.

b. Astronomical.

c. Gyro-driven.

d. Ground-type.

20. On the astro-compass, the scales for north latitude are colored

a. Red.

b. White.

c. Blue.

d. Green.

21. The astro-compass sighting assembly can be tilted for declination degrees north to degrees south.

a. 44 to 44.

b. 54 to 54.

c. 64 to 64.

d. 74 to 74.

22. The two main parts of the astrograph are a mounting ring, and

23. When installing an astrograph, be sure the mounting ring assembly is to the navigator's table.

a. At right angles.

b. At a 30° angle.

c. Parallel.

d. Close.

24. Before flight all "bubble" type sextants should be checked for -

CHAPTER

TESTING AND MAINTENANCE AIRCRAFT PLUMBING

Fittings are used to make proper connections between pressure tubing lines and aircraft instruments. However, the use of pressure tubing and hose in instruments is fast being replaced by the use of electrical remote indicating instruments. The two types of tubing used are rigid and flexible and they are used in many different sizes. Sizing is usually determined by outside diameter.

For rigid tubing there are two types of joints used at pressure tube connections. The beaded or upset type of joint is used in the vacuum, de-icer, fuel, and oil systems where hose connections are utilized. All other joints are of the flared type and are to be used in conjunction with tube connector fittings. Grip dies and flaring or beading tools are required to form flared and beaded type joints. Damaged piping and lines should be replaced with new parts. The length of tubing removed will be controlled by the location of the damage, the extent of the damage, and the most convenient location for tool manipulation. When flaring tubing, don't forget to place the fittings on tubes before flaring the ends. The flaring tool is used for forming flared tube joints. The flared type grip die used consists of two steel blocks placed side by side and held in alinement by three steel pilot pins pressed into one block and extending into corresponding holes in the other block. A number of countersunk holes are drilled along the grip die section of the tool, the holes varying in size to correspond to the piping tube sizes. The flaring tools for this type of joint consist of cylindrical bars tapered at one end to correspond to the

angles of the countersunk holes in the die. The piping tube should extend upward and through the die to approximately one-half the outside diameter of the tube. The tapered end of the flaring tool should be inserted in the tube and tapped lightly with a hammer until the walls of the tube are forced to assume the shape of the countersunk hole in the die.

To make a beaded type joint, a separate grip die is used for each size tube. The grip die consists of two steel blocks placed side by side and held in alinement by suitable pilot pins. Holes, equal in diameter to various tube sizes, are drilled through the tube die and are chamfered slightly around the top. The beading tool consists of a cylindrical rod with a chamfered hole equal to the outside diameter of the piping tube drilled approximately one-quarter inch into the end of the center. In the center of this hole, and equal in diameter to the inside diameter of the piping tube, another hole is drilled to a depth of 1" inch. A steel pilot is driven into this hole and allowed to extend 1 inch. To make a beaded joint, the piping tube is inserted into the die and allowed to extend upward from the die for approximately iy 2 diameters. The assembly is secured in a vise and the pilot of the beading tool is inserted into the open end of the piping tube. It is then hammered lightly until the desired bead is formed.

There are various types of fitting connections for use on rigid tubing. The correct size, type of coupling, and fitting should be determined to prevent further damage by a misfit and to insure a leakproof system.

There is a tendency to overtighten tubing nuts to insure that high pressure fluid will not escape. Such overtightening may severely damage or completely cut off the tube flare. If upon the removal of a tube a flare is found to retain less than 50 percent of its original wall thickness, it should be rejected.

In bending tubing, care must be exercised to prevent the collapsing of the tube at the bend and thus the restricting of flow. Bending may best be accomplished by using the bending tool. All aluminum, aluminum-alloy, or dural tub-

ing requiring bending should be bent by the bending machine. Most aluminum-alloy tubing must be annealed before bending. Do not use torch or flame on tubing because excess heat will destroy the strength of heat-treated tubing.

In making bends for fluid tubing, the bend radii as shown in table II should be adhered to.

Table II.—••nd radii for fluid tubing

Repair of flexible hose so that it may be used without ordering new fittings is not feasible. Swaging would be necessary to insert the hose into the fittings. Then the repairs would have to be tested under pressure and the machinery required is so extensive that it is impractical.

Weatherhead fittings for flexible hose lines are detachable and may be reused. The same diameter hose or fittings should be used as a replacement Weatherhead fittings may be used for lines conducting gasoline, hydraulic fluid, glycol, water, de-icing fluid, fire extinguishing medium, and air. No special tools are required in making replacements. A vise, a wrench to fit the hex nut, and a screwdriver or pocket knife are sufficient for purposes of assembly and disassembly. Weatherhead fittings can be used on low and medium pressure lines.

Aeroquip fittings for flexible hose lines are detachable and may be reused. The inside diameter of the fitting used is the

same as the inside diameter of the hose to which it is attached. Interchangeability is practical as long as inside diameters and the lengths of the hose are the same. Aeroquip fittings can be used for all.lines conducting gasoline, hydraulic fluid, glycol, water, de-icing fluid, fire extinguishing medium, and air. Special tools are needed for replacing these fittings. Hose of fittings may be replaced with the special tools in the Aeroquip Tool Kit. This kit includes an assembly block, seven adapters, two cleaning brushes, and seven assembly tools. In making an assembly, the proper size hose, end fittings, and the proper assembly tool must be used.

A proof test after assembly of Aeroquip hose and fittings must be made. To conduct a proof test, plug or cap one end of the hose and apply pressure to the inside of the hose with a hydraulic pump. The proof test pressure should be held for at least 30 seconds.

Aeroquip hose and fittings may be used on all 1000 p.s.i. lines and some lines up to 1500 p.s.i. but should not be used on high pressure lines.

Hose may be identified by the code markings. The type of hose is usually indicated by one or more solid or broken red, white, or yellow lines braided continuously throughout the length of the hose. The manufacturer's name or trademark, or a coded dot and dash strip identifies the manufacturer. The quarter of year and year of manufacture (expressed in numerals) are placed immediately above or interspersed along the identification strip.

In general, replacement of flexible hose shall be made with hose or hose assemblies of the same size, length, and design as the original parts. Hose assemblies shall be installed with sufficient length so that longitudinal stresses are not imposed. Hose contracts in length and expands in diameter when pressurized. Also, hose should be installed so that the identification strip is straight after installation. A twisted hose may fail or the attaching fittings loosen when pressure is applied. Hose installed through openings in the structure shall be protected against chafing by means of suitable grom-mets or supports. Hose and lines should be supported and held in place with suitable clamps, standards f airleads, and blocks. These supports should be placed at approximately 18-inch intervals, except in cases where the use of a support will interfere with proper performance of flexible conditions. Supports will reduce line vibration artd should be tightened enough to hold lines firmly. These supports are, in turn, bolted or bonded to the airplane.

All hose shall be visually inspected at each 30-hour check. Replacement shall be effected whenever deterioration is evident or where doubt exists as to the condition of the hose. Deterioration may be observed as ply separation, deep permanent impressions adjacent to the hose clamps, cracks in the cover revealing inner fabric, hardening, and lack of flexibility. Hose that has collapsed, in bends or due to misalinement, shall be replaced whenever its serviceability is questioned. Weather checking may be inhibited by the application of Buna-Vinylite lacquer applied to the ends and exterior of the hose.

All hose are manufactured from materials subject to deterioration by exposure to heat, sunlight, excessive moisture, and ozone. Accordingly, hose should be stored in a cool dry place and away from electrical equipment. Hose shall be stored in straight lengths to prevent set in a curved position.

In some cases locations of leaks are very difficult to determine, particularly in cases involving oxygen, gas, and air lines. To locate leaks in the above lines, soapy solutions should be used on joints around shafts, seals, or casting body. Then watch for bubbles. Generally most leaks occur because of the following conditions:

1. Flaring is poor, cracked, rough, or split.
2. Fittings are not tight enough.
3. Careless assembling.

4. Parts were not properly cleaned before assembly.

5. Crossed threads.

6. Threads damaged.

7. Wrong size or improper gasket material.

8. Not enough support.

9. Mismated parts.

DIAL PAINT

•

The requirement for radioactive luminous materials on aircraft instrument dials has been eliminated. The application of radioactive luminous material during instrument overhaul and manufacture has been discontinued. This action tends to standardize aircraft instrument dial graduations; however, differences still exist between the basic flight instruments (compass, airspeed indicator, rate of climb, altimeter, directional and horizon indicators) used for normal operations and those used specifically for night operations. To insure that each group is properly identified, the use of the present stock numbers must, of necessity, be continued. The dial presentations which are currently used are shown in table III, and it should be noted that all instruments will now have, generally, the same presentation except for the secondary scale markings of the basic flight instruments used in aircraft designated for normal operations.

Table III.—Initrument dial marking*

It should be noted that only authorized instrument overhaul activities are authorized to accomplish dial refinishing.

LIGHTING

There are two basic methods used to make instrument dial markings visible in darkness to the pilot or -crew members. The instrument face may be lighted by a lamp within the instrument case or by light from an external source, or the

principal markings and pointers may be made visible through the use of a luminescent paint The Navy does not use individually lighted instruments. Navy dial markings and pointers may have a white finish, a natural aluminum finish, or may be painted with a flourescent material. The flourescent material glows under ultraviolet light.

Another and new type of lighting, red indirect instrument lighting, has been developed and is being used by the Navy. Red indirect lighting is particularly adaptable for night vision. This type of lighting consists of lamps mounted on the instrument cover panel and enveloped by a red plastic cover. These lamps are 6 volt, 0.2 ampere, and have a midget-flange base. The lamp sockets are mounted on the cover panel and are designed so that the lamps may be replaced without removing the cover panel. The entire front surface of the instrument mounting panel is painted with nonspecu-lar black paint to insure only instrument dial reflection in darkness.

LUBRICATION OF AIRCRAFT INSTRUMENTS

Overhaul operations for aircraft instruments are to be performed by qualified personnel only. Generally instruments will require little or no lubrication in the field. The shafts and bearings of instruments are lubricated before assembly and no further lubrication should be required until the instrument is overhauled.

The Bureau of Aeronautics has issued instructions regarding the approved lubricants which are applicable to all aircraft instruments requiring lubrication. These instructions apply only to overhaul activities when processing aircraft instruments requiring overhaul or lubrication. Table IV, page 228, is a list of the approved lubricants for aircraft instruments and their specific application.

RANGE MARKING

Range markers have sometimes been applied to instrument cover glasses by operating units to indicate operating ranges and maximum allowable limits in order to facilitate check-

ing instrument readings. This is considered to be desirable provided a uniform system of marking is followed.

In aircraft which have ultraviolet or white instrument lighting, green arcs are used to indicate operating ranges and red strips to indicate limits. The standard stock markers are fluorescent and the arcs appear green and the red strips red when activated by ultraviolet or viewed under white light. In aircraft which have red lighting, white arcs are used to indicate operating ranges and red strips are used to

indicate limits. Under red lighting, both the white arcs and the red strips will appear red.

The range markers are applied on the instrument cover glass. The glass must fit snugly since any movement of the cover glass will result in incorrect range markings. To insure that any such movement would be readily discernible, a short white radial line may be painted on the bezel so that it extends onto the cover glass.

INTERCHANGEABLY OF AIRCRAFT INSTRUMENTS

Extreme care must be exercised in the substitution of aircraft instruments. The Erection and Maintenance Manual for the type aircraft should be checked for the correct type of instrument. This manual usually lists the stock number of the instrument. In many cases when replaceable instruments are available, then substitution may prevent undue delay in returning aircraft to service. Aviation Supply items bearing the same stock numbers are operationally interchangeable regardless of the manufacturer or his part number. An instrument which has been assigned a stock number has at least one distinctive feature which has made it advisable to identify it individually. Normally one instrument will not be substituted for another. Before substitution is effected, the instrument should be properly identified in the Class 88 Section of the Aviation Supply Office Catalog, the Erection and Maintenance Manual, and pertinent Technical Orders.

TROUBLE SHOOTING

The operation of most aircraft instruments is entirely automatic. Once installed, the units require no further maintenance or servicing, other than the performance of routine and periodic inspections. If malfunctioning of a system or an instrument occurs, it is first necessary to localize the source of trouble. A systematic trouble shooting procedure, including the possible service troubles and their remedies, should be developed for each type of instrument. Much of this information may be found in pertinent Aeronautical

publications and Erection and Maintenance Manuals for specific type aircraft instruments and for aircraft.

An instrument reported as functioning improperly, or otherwise suspected of being unserviceable, must first be checked to determine whether the instrument or the installation is at fault. Usually troubles fall into three groups: trouble in the power supply; trouble in the unit; or trouble in the connections to units, either electrical or mechanical If the installation is faulty, it can be corrected by line maintenance. This is usually found to be trouble in the power supply. If the instrument is the cause of the trouble, in most cases it must be removed and replaced with a serviceable unit The defective unit then should be sent to a qualified instrument overhaul depot for detailed inspection, overhaul, and repair.

To check the power source of electrical instruments, use the voltmeter on the instrument panel; however, if a voltmeter is not provided, an appropriate a-c voltmeter should bfi used. If no frequency meter is available, the frequency output from the inverter should be checked. Even though correct power is being furnished, it may fail to reach the unit. Therefore, the voltage from the power supply to the unit suspected of malfunctioning should also be checked across the power leads of the electrical connector plug going to the suspected unit.

The power supply should be disconnected by means of the switch in the pilot's cockpit before removing any electrical connector plug or making any repairs near the electrical contacts of the unit. Turning off the power eliminates the possibility of accidental spark ignition of gasoline vapors.

When making ground tests of electrical instruments, an external power supply should be connected to the aircraft's external battery connections. The aircraft battery should not be used in conducting ground tests of equipment. An auxiliary power unit, electrical converter, or engine converter should be used to provide a source of external power.

Many of the electrical instruments, Magnesyn indicators, for example, may be checked by use of a Magnesyn test in-dicator to determine whether the trouble lies in the unit or in its electrical connections. There are other test instruments that may be used to determine the condition of a unit that will be discussed later.

In checking the electrical circuit, an ohmmeter should be used, or where applicable, a suitable voltmeter. The use of a continuity tester will prove to be invaluable in many cases; however, its use on some circuits is not recommended. For example, in checking the connections between a Magnesyn transmitter and an indicator, a defective circuit caused by a high-resistance, improperly made soldered connection will not be indicated on a continuity tester. On an ohmmeter, however, it will be indicated in the form of a much higher resistance reading than is normal for the circuit in question. All wiring should be checked for continuity. This is important especially in Magnesyn circuits as a Magnesyn may function if one of the lead wires is broken although the indications may not be correct.

Particular attention should be given to the electrical connector plugs because it is easy for inexperienced personnel to dismantle the plug instead of turning the coupling nut and then pulling the plug.

TEST EQUIPMENT

Test equipment should be handled with care at all times, as the slightest bump or jar can result in damage. Proper procedure in the use of test equipment should be carefully followed, too. Improper use of testers will not only damage the instruments, but also the testers themselves. Such practices may also lead to injury of the personnel using the testers.

Testers should be stowed in their proper places, and in neat, orderly arrangement at all times, so that they can be easily reached and used. This orderly manner of arranging testers contributes greatly not only to appearance, but also to the general efficiency of the shop. All testers should be kept free from moisture, dirt, and excessive oil.

MA6NESYN TEST TRANSMITTER

The Magnesyn Test Transmitter is designed as a service standard against which to calibrate single indicators manufactured by the Eclipse-Pioneer Company. Comparison of the degree of rotation set in through the drive mechanism of the test transmitter with the reading on the Magnesyn indicator enables the operator to determine the degree of error of the indicator.

The Magnesyn Test Transmitter consists essentially of a Magnesyn transmitter assembly, a compensated cam drive assembly, and a base assembly. The Magnesyn transmitter may be rotated through 360°, by means of the cam and gear drive, to transmit to the instrument under test the desired degree of rotation. A self-contained d-c supply enclosed in the transmitter case provides correctly polarized voltage for an electrical zero check of the instrument under check. To set the transmitter for operation, the switch knob should be moved to off and the power cord connected to a suitable 26-volt, 400-cycle, single-phase cycle.

This test transmitter is a rugged instrument and needs little maintenance and inspection for service. Every 30 days the drive mechanism should be checked for smooth rotation through 360°. The switch should rotate without forcing and all cables and connectors must be in good working condition.

PORTABLE FIELD TEST SET

Instrument shops will have test sets to aid in testing, servicing and maintenance of instruments. The portable instrument field test set is a small laboratory mounted on a portable three-wheel carriage. It is an assembly of precision mechanical devices and instruments. Its design and construction permit testing and inspection of various aircraft instruments without removing them from the airplane. It is especially suited for use on aircraft carriers and is sufficiently light and compact to be carried by a cargo plane or truck.

There are any number of sets available. You should become familiar with your set and know its limitations. Persons responsible for the working order of the test set should refer to the book of instructions that is supplied with the set.

EMERGENCY PROTECTIVE TREATMENT FOR AIRCRAFT INSTRUMENTS

Aircraft instruments that have been submerged in water will as soon as practicable after submergency, be given treatment to minimize corrosion of parts.

The instrument case should be opened and the mechanism disassembled to the extent warranted for the particular type of instrument in question. All parts of the mechanism then are flushed thoroughly with a water-displacing, rust preventative compound. The flushing action may be carried out in a 3 or 4 gallon container, or the instrument may be filled with the protective compound in any convenient manner. The instrument should be shaken vigorously to insure thorough contact of the protective compound with all parts of the mechanism. The instrument is then emptied of liquid so that water will not be allowed to settle in the bottom of the instrument case. This operation is repeated at least twice. After treatment, the mechanism will be allowed to drain and dry. The instrument, with all parts which have been removed, should be packed carefully and shipped to the overhaul base for immediate repair. The instrument is tagged and the shipment marked externally, "Immersed in water."

The flushing compound may be poured into a container and allowed to settle. The part that does not contain water should be returned to the original storage container for further use.

Gyroscopic instruments, such as horizons, directional gyros, turn and bank indicators, automatic pilot horizon control units and automatic pilot directional control units are partially disassembled prior to protective treatment in order to insure contact between the flushing compound and the more inaccessible parts of the mechanism during the flushing operation.

In general, instruments should be disassembled to the extent that the various parts of the mechanism will be thoroughly flushed. For treating some manifold pressure gages, it will be necessary to remove the back cover in order for the rust preventing compound to have free access to the aneroid compartment.

REPAIRS

The maintenance and major repair or overhaul of Class 88 aircraft instruments is accomplished in accordance with Bu Aer instructions. "Maintenance*' includes the proper handling and storage of instruments, inspection of instruments, inspection of instruments drawn from stock, periodic inspection of instruments installed in aircraft, to insure that only instruments in a satisfactory operating condition are in use, and all other work of a maintenance nature which will insure that the best possible service is being obtained from the instruments.

The terms "major repair" and "overhaul" include all work which requires disassembly of the instrument or replacement or adjustment of any part of the instrument mechanism inside the case. Major repair overhaul of Class 88 aircraft instruments is done only by those activities so authorized by the Bureau of Aeronautics.

SHIPPING AND HANDLING

Instruments in need of repair should be handled with the same care that is exercised in handling new instruments, in order that additional damage due to improper handling may be avoided. In order to prevent damage in handling and storage, all instruments insofar as practicable are stored and transported in individual cartons. Care should be taken in packing damaged instruments to insure that no additional damage will result during shipment. Individual cartons are packed for shipment in strong wooden boxes, except when the means of shipment, such as by air, prohibits such packing.

Instruments requiring caging, such as fluxgate compass transmitters, directional gyros, and gyro-horizon indicators should be placed in the caged position prior to shipment.

Refer to manufacturers' instructions for detailed information regarding caging.

The information in the following paragraphs outlines the packaging operations to be used in packaging aircraft instruments for shipping. The instructions regarding cleaning, wrapping, use of cartons and use of labels apply to all instruments. Those regarding the use of cushioning material are supplemented in Table V by specific information concerning the limiting thicknesses of this material for various groups of instruments.

Cleaning

All instruments should be thoroughly cleaned, prior to packaging, in such a manner that they will be ready for immediate use after unpacking. Extreme care should be exercised to prevent contamination of instruments with perspiration and to thoroughly dry the instruments before wrapping with moisture impervious material.

Cartons

When packaging instruments, use fiber corrugated cartons of the regular slotted type. Different sizes are available. Select the size which will allow the use of at least the minimum thickness of cushioning material required. Never use too small a carton—one that will not permit the inclusion of the proper amount of cushioning material.

Labeling

You, as an instrumentman, will see small decalcomania on aircraft instruments when you are assigned to an activity which installs and removes instruments from aircraft. A locally prepared decalcomania should be placed on all instruments in such a position as not to interfere with proper operation. This will provide a place to record overhaul, installation and removal data. All records of previous overhaul are to be removed at the time of each new overhaul. Information as to overhaul activity and date overhauled is to be filled in by the overhaul activity. The installation activity will fill in information regarding installation and removal.

Cushioning

For cushioning the instruments when packaging, the best material is dry cellulose crepe wadding. Be sure it is dry. Also, be sure it is of sufficient thickness to meet the requirements given in table V, which lists a number of aircraft instruments and the limiting thicknesses of cushioning material which must be used with each. The instruments are coded with a group number, the meaning of which is given at the end of the table. The outside layer and any extra material required to fill up the carton should be accordian folded. Pack the carton firmly to prevent motion of the instrument and resulting damage during transit.

Table V.—Cushioning material limit*

INSTRUMENT NAME	PACKAGING GROUP CODE LETTER
Accelerometer	B
Airspeed Indicator	B
Altimeter	B
Anti-Icing Fluid Qty. Gage	B
Attitude Gyro	D
Autopilot, Air Hydraulic	
(Sperry and Jack & Heintz)	
Bank and Climb Control	D
Directional Control	D
Mounting Unit	B
Proportional Bank Adapter	BB
Rudder Control	B
Servo	B
Speed Control Valve „	B
Drain Trap	A
Filters	A
Pressure Regulators	A
Transfer Valve	B
Sump Tank	A
On-Off Control	A
Auto Pilot Electric (Pioneer P-l)	
Relay Box	B
Controller	B
Gaging Relay	B
Master Directional Indicator	D

INSTRUMENT NAME	PACKAGING GROUP CODE LETTER
Auto Pilot Electric—Continued	
Fluxgate Transmitter	D
Servo Disconnect	B
Fluxgate Transmitter Amplifier	C
Bank and Turn Indicator	C
Gyro Horizon Indicator	D
Amplifier Adapter	A
Power Junction Box	B

Servo B
Servo Amplifier C
Clutch Switch B
Auto Pilot, Electric Hydraulic (Grumman GR-1) (General
Electric G-l)
Dir. Gyro Control D
B & C Control D
Followup Control D
Servo Amplifier C
Junction Box B
Transfer Valve B
Hydraulic Servo B
Oil Pressure Regulator A
Drain Filter A
Oil Filter A
Clock Aircraft C
Compass, Astro B
Compass, Magnetic, Standby C
De-Icing Pressure Gage B
Engine Gage Unit B
Flow Meter Transmitter B
Flow Meter Indicator B
Fuel Pressure Gage B
Fuel Pressure Transmitter B
Fuel Quantity Indicator B
Fuel Quantity Transmitter B
Gyro Directional D
Gyro Fluxgate Amplifier C
Gyro Fluxgate Caging Control B
Gyro Fluxgate Caging Switch B
Gyro Fluxgate Indicator D
Gyro Fluxgate Transmitter D
Hydraulic Pressure Gage _,, B
Inclinometer B
Manifold Pressure Gage (Direct) B
Manifold Pressure Indicator B
Manifold Pressure Transmitter B
INSTRUMENT NAME PACKAGING GROUP CODE LETTER
Clock Aircraft—Continued
Oil Pressure Gage B
Oil Pressure Indicator B
Oil Pressure Transmitter B
Oil Quantity Indicator B
Position Indicator B
Position Transmitter B
Pitot Static Tube A

Rate of Climb Indicator B
Resistance Therm. Ind B
Suction Gage B
Synchroscope B
Tachometer Indicator B
Tachometer Generator B
Thermocouple Indicator B
Turn and Bank Indicator C
Code: Required Thickness of
Packaging Group Cushioning Material
Use only enough to fill out carton A
1 inch (-% + %) B
2 inches (-%+%) C
3inches (-%+!> D

INSTRUMENT INSPECTION FORM

When you are assigned to the operating Fleet, you will find that a detailed instrument checkoff list must be used in conjunction with routine aircraft checks. The following sample list, table VI, may be used as a guide toward- the formulation of a detailed instrument checkoff list adaptable to local conditions.

Tabic Vl.^~lntfitinicfit inspection form
3 »
ITEM | S % g g \'7bJ O
NO. NATURE OP INSPECTION pu, < Q g § § 3
1. Check normal position of pointers on all indicating instruments I I I I I I I
2. Check cover glasses for cracks, looseness and cleanliness I I I I I
3. Check condition of instrument dials __ I I III
4. Visually check pitot-static tubes for ob-structions and alinement I I I
5. Wind clock C
6. Inspect quantity gage systems for correct indications I I I ,,
7. Ground check autopilot for operation and overpower __ C ,, __ __
8. Ground check gyro fluxgate compass for operation ... C ..
9. Clean and replace gyro air filters __ __ RRRR
10. Drain pitot-static tubes and clear lines. _ RRRR
11. Clean and replace autopilot oil filter ...... RRRR
12. Check inverter brushes and commutator. RRRR
13. Check instrument panel installation and panel mounting of all instruments _____ C C C C
14. Check mounting of all transmitters and autopilot components __ .. .. C C C C
15. Check mounting of gryo fluxgate compass

components C C C C

16. Check synchronization of gyro fluxgate compass caging control .. .. ._ __ .. C

17. Perform complete functional tests on all indicators and transmitters .. __ __ __ C

18. Bench test all gyro instruments including autopilot control units __ __ __ R

19. Bench test gyro fluxgate transmitter and indicator.. R

20. Perform functional tests on autopilot system .. _. ._ C

21. Perform functional tests on gyro fluxgate compass system .. __ C

22. Check liquid filled compass indicators and transmitters for correct liquid level I

23. Check inverter output . . .. ._ . .. C

24. Clean all external electrical connections and check receptacles for corrosion ._ ._ .. __ __ C

25. Check all instrument air, vacuum, and hydraulic systems for leaks C

26. Check external electrical systems for continuity and resistance - . _. __ . - C

27. Check and regulate gyro vacuum system. .. ._ _. -- C

28. Check and regulate autopilot oil pressure. .. .. ..-- C

29. Compensate the compass installations... .. .. .. -- C

Code:

I—Inspect (visual only).

C—Check (functional).

R—Remove, recondition or replace.

QUIZ

1. What are the two types of joints used for rigid tubing?

2. What is the maximum pressure allowed for use with Aeroquip hose?

3. Hose contracts in and expands in with pressurized.

4. What is the best means to locate leaks in oxygen, gas, and air lines?

5. When range markers are applied on the instrument cover glass, the glass must fit snugly. Why?

6. What is the first step in trouble shooting if malfunctioning of a system or an instrument occurs?

7. Who shall perform major repair or overhaul of Class 88 aircraft instruments?

8. All instrument air, vacuum, and hydraulic systems should be checked for leaks at the hour check.

9. A proof test after assembly of Aeroquip hose and fittings must be made. How is pressure applied?

10. The use of pressure tubing and hose on instruments is fast being replaced. By what?

11. When bending tubing, what is the most likely trouble which might be encountered?

12. How can the age of hose be determined?

13. Why should hose be stored in a cool dry place and away from electrical equipment?

14. Generally, lubrication of aircraft instruments is performed

15. Normally, one aircraft instrument will not be substituted for another; however, items bearing the same stock numbers are

16. What instruments are commonly used to check the power source of electrical instruments?

17. What is the best cushioning material for packing instruments?

18. What type of pressure lines may be used with Weatherhead fittings?

19. Hose is identified by code markings. How is a manufacturer's name or trademark coded?

20. What typ* of dial paint is used for primary graduations and numerals on basic flight Instruments?

21. Red indirect instrument lighting is particularly adaptable for

22. What is the danger of overtightening tubing nuts?

23. Why should hose be installed so that the identification strip is straight after installation.

24. When range markers are installed on the cover glass of instruments, what means are used to discern any movement of the cover glass?

25. Usually troubles with aircraft instruments fall into three groups. List them.

26. Ail work which requires disassembly of the instrument or replacement or adjustment of any part of the instrument mechanism inside the case shall be termed as

27. Why is it not feasible to repair flexible hose?

28. List three causes of leaks in hose.

APPENDIX I

ANSWERS TO QUIZZES

CHAPTER 1

INTRODUCTION TO INSTRUMENTS

1. Engine instruments.

2. Flight

3. Navigation.

4. D.

5. Not acceptable.

6. Periodic functional testing.

7. Fuel or lubrication, a C.

0. Careful handling.

CHAPTER 2

PRESSURE

1. Force.

2. Force.

3. Inches of mercury.

4. 14.7

5. 29.92 inches.

6. Barometer.

7. B.

8. Sealed airtight.

9. B.

10. Incompressible. Compressed.

11. Bourdon.

12. Pounds.

13. A.

14. Diaphragm capsule.

15. Rate-of-climb.

16. Mechanically, electrically, vacuum, or by gyro.

17. 59.

18. Inversely as the volume.

19. Metal tubing.

CHAPTER 3

REMOTE READING INSTRUMENTS

1. D-C selsyn.

2. Indicating element.

3. Circular resistance.

5. Pewer.

6. Polarity.

7. Transmitter and an indicator.

8. Rotor. Stator.

9. Duplication of motion.

10. Engine functions.

11. Indicator. Transmitter.

12. Two-wire.

13. Polarity.

14. Stator.

15. Stator winding.

16. Engine driven alternator.

17. D.

18. Magnesyn.

19. Two-wire.

20. B.

21. Rotor.

22. A.

23. Two individual.

24. Indicator.

25. Toroidal.

CHAPTER 4

PRESSURE GAGES

1. Bourdon.

2. Pressure leaks.

3. Electrical transmission.

4. Pointer. Graduated dial.

5. 0 to 25.

6. D.

7. Air-driven gyroscopic.

8. Antiseize compound.
9. C.
10. Inches of mercury.
11. Carburetor.
12. One-half inch yellow stripe.
13. A.
14. Jewel pivots.
15. Every 1000 hours.
16. 10.
17. 15-17.
18. D.
19. Manifold pressure.
20. The same as local barometric pressure.
21. C.
22. Conform.
23. 0to25.
24. D.
25. Suction.
26. Oscillation.
CHAPTER 5
AIR PRESSURE MEASURING INSTRUMENTS
1. C.
2. Quiet-undisturbed,
3. Tap the instrument panel.
4. Installation. Poor adjustments.
5. Measure altitude above some fixed point on the ground.
6. B.
7. On zero.
8. Scale correction cards.
9. Rate-of-climb indicator.
10. C.
11. Pitot
12. Union connections.
13. B.
14. Two.
15. B.
16. Removed for overhaul.
17. C.
18. 6000.
19. Use the zero adjusting knob.
20. Brass, copper.
21. Covered with a sack of leather or cloth.
22. 28.1 to 31.0.
23. Obtain the reading from the station altimeter.
25. A.
26. A leak in the static line

CHAPTER 6
THERMOMETERS
1. Metals.
2. More.
3. Reflect heat from other sources than the air.
4. Bulb, capillary tube, indicator.
5. Properly labeled.
6. Indicating instrument, the bulb.
7. Wbeatstone's.
8. Loose wiring connections.
9. Two wires of different metals.
10. Copper gasket.
11. D.
12. Copper.
13. B.
14. Resistor. Resistances.
15. Copper. Constantan.
16. Cockpit.
17. Millivoltmeter.
18. Minus 40 to plus 40.
19. A.
20. Vapor pressure.
21. Heat-sensitive resistor.
22. Clean. Tight.
CHAPTER 7
TACHOMETERS
1. Engine crank shaft.
2. Indicator and a generator.
3. Electric cable.
4. Universal.
5. Turning at the same rate.
6. Master engine.
7. Tachometer generators.
& B.
9. Three individual instruments.
10. Slippage.
11. Interchangeable.
12. Stud. Screw.
13. Decrease the strength of its magnets.
14. Does not run.
15. Master.
16. Dual.
17. 35,000.
18. Some additional support
19. Four-engine.
20. Dual.

CHAPTER 8
FUEL GAGES AND FLOWMETERS
L A rotor, element coils, and a cone.
2. Coil fields, rotor magnet.
3. Bar magnet
4. Empty.
5. Free from burrs or metal protrusions.
6. Strikes the top and bottom of the tank.
7. Autosyn fuel flowmeter.
8. No.
9. Relief valve.
10. Pressure or gravity-feed.
11. Small bar-magnet.
12. Counterclockwise.
13. D C selsyn.
14. Float-and-arm. Direct lift float
15. B.
16. 28Voltd-c.
17. Ratiometer.
18. Has no effect on the readings.
19. Damping ring of copper.
20. C.
21. Can register its complete range.
22. Operate without electrical interference.
23. A dial-change indicator.
24. Electronic. Pounds.
25. Condenser.
CHAPTER 9
COMPASSES
1. Standby compass.
2. An expansion chamber.
3. Dampen the oscillations.
4. Two.
5. Nonmagnetic metal.
6. Compensators.
7. D.
8. Transmitter. Indicators.
9. A master indicator, repeaters, an amplifier, a remote caging unit.
10. Clockwise.
11. Expansion and contraction of the liquid.
12. Specifically authorized.
13. D.
14. Clear and at the proper level.
15. Plus.
16. C.
17. A.

18. Amplifier.
19. 90.
20. Mounted on instrument panel. Mounted face upwards.
21. D.
22. Secured where it would be in normal flight.
23. B.
24. Minus.
25. B.
26. Magnesyn.
27. Gyro stabilized flux gate.
28. Five.
CHAPTER 10
GYROSCOPES
1. Remain in a fixed direction in space.
2. Resist.
3. A.
4. Erection.
5. A stream of air.
6. Venturi tube.
7. Turn and bank.
8. Two.
9. Withdrawing.
10. Eight.
11. At right angles.
12. Lateral attitude.
13. Bank.
14. Dash pot.
15. A poor electrical connection.
16. Cage.
17. D.
18. C.
19. Universally.
20. B.
21. Flow of air passing the aircraft.
22. Proportional. Speed.
23. An inclinometer.
24. B.
25. O.
26. 55.
27. B.
28. Horizontal.
CHAPTER 11
AUTOMATIC PILOTS
L Two gyroscopes.
2. Vertical.
3. Is returning to normal position.

4. Servo relief valves.
5. B.
6. Two.
7. Hard over to hard over.
8. Stick-type controller.
9. (a) 70, (t» 70, (c) 70, (d) 360.
10. Two pendulums and two torque motors.
11. Speed control valves.
12. Directional gyro control.
13. Can be raised or lowered in front of the horizon bar.
14. Return line.
15. Clutch switch.
16. Compensation.
17- Magnetic amplifier. Control amplifier.
18. Solenoid operated.
19. Magnetic heading of the plane.
20. Position-type.
21. B.
22. Bypass.
23. Rudder knob. Instrument dial.
24. Hydraulic cylinders.
25. Autosyn.
26. Synchronized.
27. A.
28. Remote compass transmitter.
29. Constant barometric flight.
30. C.

CHAPTER 12
NAVIGATION INSTRUMENTS
1. Wind on your aircraft.
2. Relative to various land-marks.
3. Heading. Track.
4. Astro-compass.
5. Fore-and-aft.
6. Curved lines.
7. D.
8. Sextant.
9. At its lowest point of traveL
10. B.
11. Open vane-type sights.
12. Daily.
13. D.
14. A.
15. D.
16. Bubble.
17. Astigmatizer.

18. Smoke light or bomb.

19. 380.

20. B.

21. B.

22. C.

23. Detachable projector.

24. O.

25. Bubble forming ability.

CHAPTER 13

TESTING AND MAINTENANCE

1. Beaded and flared.

2. 1500 p.s.i.

3. Length. Diameter.

4. Soapy solutions should be placed on joints and seals. In case the solution bubbles, a leak is present

5. Any movement of the cover glass will result in inaccurate range markings.

6. Localize the source of trouble.

7. Only those activities so authorized by the Bureau of Aeronautics.

8. 120 hour.

9. By use of a hydraulic pump.

10. Electrical remote indicating instruments.

11. Care must be exercised to prevent collapsing of the tube at the bend.

12. The quarter of year and year of manufacture (expressed in numerals) are placed immediately above or interspersed along the identification strip.

13. Hose is manufactured from materials subject to deterioration by exposure to heat, sunlight, excessive moisture, and ozone.

14. Overhaul personnel.

15. Interchangeable.

16. Voltmeter—a-c or d-c, frequency meter, and ohmmeter.

17. Cellulose crepe wadding.

18. Low and medium pressure lines.

19. The manufacturer's name or trademark or a coded dot and dash strip identifies the manufacturer.

20. Yellow fluorescent.

21. Night vision.

22. Overtightening may damage or cut off the tube flare.

23. A twisted hose may fail or the attached fittings loosen when pressure is applied.

24. A short white radial line may be painted on the bezel so that it extends onto the cover glass.

25. 1. Trouble in the power supply.

2. Trouble in the unit.

3. Trouble in the connections to units.

26. Major repair or overhaul.

27. Swaging is necessary. Then the hose would have to be tested under pressure. The machinery required is so extensive that it is impractical.

28. 1. Fittings are not tight enough.

2. Crossed threads.

3. Wrong size or improper gasket material.

APPENDIX II

QUALIFICATIONS FOR ADVANCEMENT IN RATING AVIATION ELECTRICIAN'S MATES (AE)

Rating Code No. 6800

General Service Rating Scope

Aviation electrician's mates maintain, adjust, test, repair, and replace all aircraft electrical power generating and converting, lighting, control, and indicating systems and components. Inspect, maintain, and install all aircraft electrical wiring. Maintain, adjust, test, and replace aircraft flight and engine instrument systems.

Emergency Service Ratings

Aviation Electrician's Mates M (Electricians),

Rating Code No. 6801 AEM

Aviation electrician's mates maintain, adjust, test, repair, and replace all aircraft electrical power, lighting, control, and indicating systems and their components. Inspect, maintain, and install all aircraft electrical wiring.

Aviation Electrician's Mates 1 (Instrument

Repairmen), Rating Code No. 6802 AEI

Maintain, adjust, test, and replace aircraft electrical flight and engine instrument systems and mechanical and vacuum instruments.

Navy Job Classifications and Codes

For specific Navy job classifications included within this rating and the applicable job codes, see Manual of Enlisted Navy Job Classifications, NavPers 15105 (Revised), codes AE-7100 to AE-7199.

Qualifications for Advancement in Rating

Applicable Rates

Qualifications for Advancement in Rating ——— —— —

100 PRACTICAL FACTORS

101 Operational

1. Demonstrate method of resuscitating a man unconscious from electrical shock and of treating for electrical burns 3 3 3

2. Observe applicable safety precautions while working in or about aircraft and those prescribed for shop and line electrical maintenance. 3 3 3

3. Operate Ground Electric Power Units required

in the service and maintenance of aircraft 3 3 3

4. Perform operational test of aircraft electrical systems, including such flight tests as required

by own activity % 2 „

»

102 Maintenance and/or Repair

1. Demonstrate safe and proper use and care of

hand and power tools common to the rating 3 3 3

2. Detect, localize, and correct faults in electric lighting and power circuits, using a multimeter in testing for continuity, short circuits, and

grounds 3 3 3

3. Read and work from schematic wiring diagrams in the maintenance and installation of aviation electrical circuits 3 3 3

4. Identify characteristics of resistors and capacitors by the RMA code. 3 3 3

5. Fabricate all types of cables used in aircraft electrical circuits, employing proper soldering
and insulating techniques 3 3 3

6. Use handbook of maintenance instructions required in the maintenance of aviation electrical
and instrument equipment 3 3 3

7. Demonstrate safe and proper use of general
test equipment furnished own unit 3 3 3

8. Check and replace circuit breakers, fuses, and bonding wires - 3 3

9. Remove and install aircraft power generating equipment, voltage regulators, and system protective devises 3 3

10. Check condition of aircraft batteries; maintain,
replace, and test them for proper charge 3 3

Qualifications for Advancement in Rating

Applicable Rates

AE AEM AEI

102 Maintenance and/or Repair— Continued

11. Disconnect and connect associated wiring to air
frame and engine accessories. 3 3

12. Install and test electrical wiring on Quick
Engine Change Units 3 3 __

13. Replace and compensate electric remote indicating and magnetic compasses 3 2 3

14. Perform preventive maintenance, including external cleaning, lubricating, checking, replacing, and making minor adjustments to mechanical,
electrical, and vacuum instruments 3 2 3

15. Perform preventive maintenance of entire aircraft electric wiring installation including wire, insulating materials, clamps, terminal strips,
and connectors 3 3 2

16. Perform preventive maintenance on automatic
pilot equipment 2 2 2

17. Make performance tests, including bench checks and adjustment of aircraft power generators, inverters, regulating controls, and power system protective devices I 2 2 _.

18. Maintain, test, and adjust electrical indicating systems, including warning systems 2 2

19. Demonstrate safe and proper use of special test equipment furnished own activity for maintenance of:

a. Aviation electrical equipment 2 2 _ _

b. Aircraft engine and flight instruments 2 2

20. Perform insulation resistance tests on wiring, using a megger or appropriate high voltage insulation tester 2 2 ,,

21. Perform periodic checks on Ground Electric
Power Units to insure proper electric output. . 2 2 _ .

22. Perform electrical tests on components of ignition systems for proper operation 2 2

23. Maintain, adjust, and perform preflight checks on aircraft searchlights and their

power supply

systems __ 2 2 „

24. Maintain, test, replace, and adjust the electric portions of aircraft electric-hydraulic systems. 2 2

25. Test, adjust, calibrate and make authorized
repairs to aircraft instruments 1 2 „

Applicable Rates

Qualifications for Advancement in Rating

AE AEM AEI

102 Maintenance and/or Repair —Continued

26. Maintain and replace electric components of aircraft cabin pressurization, air-conditioning,
and heater systems 1 1 „

27. Maintain and test aircraft propeller and engine electric/electronic control systems and their components 1 1 „

28. Make performance test including bench, pre-flight, and required in-flight adjustments to maintain proper operation of automatic pilot equipment 1 1 „

29. Test, adjust, and make authorized repairs to electronic components used in instrument, power, automatic flight, indicating, and lighting systems 1 1

30. Test, adjust, and make authorized repairs to components of aircraft electrical systems including servos, relays, protective devices, and
all rotating electric equipment... 1 1 „

31. Perform instrument repair, using required machines and special hand tools __ ' „ 1

32. Interpret wiring diagrams contained in handbooks of maintenance instructions in trouble
shooting electrical systems 1 1 C

33. Analyze malfunctions and determine corrective action required on:

a. Aircraft electrical systems C C

b. Aircraft instrument systems C C

103 Administrative and/or Clerical

1. Make required entries in aviation electrical
shop maintenance records 3 3 3

2. Use NavAer Publications Index to locate, identify, and obtain technical publications 3 3 3

3. Complete electrical instrument section of the Standard Aircraft Inventory Log 2 2 2

4. Determine part and stock numbers from available technical supply publications for obtaining replacement materials 2 2 2

5. Conduct on-the-job training and supervise personnel engaged in maintenance of:

a. Aircraft electrical systems 1 1

b. Aircraft instrument systems - 1 1

Qualification* for Advancement in Rating

Applicable Rates

AE AEM AEI

103 Administrative and/or CLERiCAL^Continued

6. Furnish technical assistance in preparation of reports required by higher authority relating to electrical systems and/or equipment, including RUDMS, RUDAE, and aircraft

accident re-

ports C C C

7. Organize and administer personnel and facilities for maintenance of aviation electrical and/or instrument systems C C C

8. Supervise the use, filing, and maintenance of publications and records; supervise preparation of reports required by own department—. C C C

9. Supervise the requisition and inventory of, and account for allowed materials in accordance

with current directives C C C

10. Screen defective exchangeable electrical components and instruments, for feasibility of authorized local repair in lieu of exchange C C C

200 EXAMINATION SUBJECTS

201 Operational

1. Effects of electrical shock and methods of artificial respiration 8 3 3

2. Safety precautions to be observed in working in

or near airplanes and on electrical equipment.. - 3 3 3

202 Maintenance and/or Repair

1. Determine values of voltage, current, and resistances in both series and parallel resistor combinations --- 3 3 3

2. Given any two of the following valuies, solve for the remaining: (a) frequency, capacitance, and capacitive reactance; (b) frequency, inductance,

and inductive reactance 3 3 3

3. Mathematical relationships between average, effective, and peak values of voltage and current in a. c. circuits - 3 3 3

4. Application of the laws of magnetism to simple

d. c. motors and generators 3 3 3

5. Operating principles and use of ohmmeters, voltmeters, and ammeters 3 3 3

6. Operating principles of and maintenance procedures for primary and secondary batteries. . . 3 3 3

Qualifications for Advancement in Rating

Applicable Rates

AE AEM AE1

202 Maintenance and/or Repair— Continued

7. Identification of standard wiring code and

wiring diagram symbols 3 3 3

8. Proper use of hand and small power tools and measuring instruments common to the rating.. 3 3 3

9. Types and characteristics of capacitors and resistors employed in aircraft electric and instrument equipment 3 3 3

10. Principles of hydraulics applicable to aircraft instruments.. 3 __ 3

11. Operating principles of gyroscopes 3.-3

12. Types and uses of aircraft a. c. and d. c. motors

and generators 3 3 2

13. Types and uses of aircraft instruments 3 2 3

14. Principles and uses of synchros in aircraft control and indicating systems including

magnesyn, autosyn, and d. c. selsyn 2 2 3

15. Calculations involved in changing range of d. c.
meters by use of shunts and multiplier resistors 2 2 2

16. Principles of operation and applications of single phase current and voltage transformers
in aircraft electrical systems 2 2 2

17. Effects of meter sensitivity in amplifier circuit
voltage measurements 2 2 2

18. Principles of operation of thermocouples as used in aircraft fire detector, temperature indicating, and control circuits 2 2 2

19. Determine values of: (a) voltage, current or capacitance in both series and parallel capacitor circuits; (b) voltage, current, or inductance
in a series inductor circuit 2 2 ,,

20. Solve problems involving relations between impedance voltages and line current in single-phase series a. c. circuits 2 2

21. Principles of electro-magnetic induction as employed in aircraft magnetos and engine ignition
starting devices 2 2 __

22. Method of operation, characteristics, and functions of reverse current, over voltage, and feeder protective devices used in aircraft d. c.
power generating systems 2 2 __

Applicable Rates

Qualifications for Advancement in Rating

AE AEM AD

202 Maintenance and/or Repair —Continued

23. Method of operation, characteristics, and maintenance techniques for finger type and carbon pile voltage regulators as used to control aircraft generators and inverters 2 2

24. Factors and calculations involved in selection
of proper wire for aircraft electric circuits 2 2 _ _

25. Methods of speed regulation used in aircraft inverters 2 2

26. Given two of the following values in a d. c. circuit, solve for the remaining: Power, voltage,
current and resistance 2 2 1

27. Voltage, current, and power relationship involved in three phase wye and delta a. c. power distribution systems 2 2 1

28. Characteristics of triode, pentode, and tetrode
vacuum tubes 8 2 1

29. Operating principles of half- and full-wave rectification 2 2 1

30. Operating principles of diode and dry-disk
rectifiers 2 2 1

31. Methods of obtaining self and fixed bias for
operation of vacuum tubes 2 2 1

32. Principles of operation of oscillator, amplifier, and signal discriminating circuits used in remote indicating compass, and automatic pilot
systems 2 2 1

33. Purpose and application of measuring instruments, including oscilloscopes, tube

testers, a. c. meters, vacuum tube voltmeters and Wheatstone bridge 1 1 1

34. Relationship of the current, voltage, and impedance in LC circuits at, above, and below resonant frequency 1 1

35. Types, internal connections, and methods of testing armatures and field windings employed in aircraft a. c. and d. c. electrical rotating equipment- 1 1

36. Principles of operating single and polyphase

a. c. motors and generators 1 1 „

Qualifications for Advancement in Rating

Applicable Rates

AE AEM AE1

202 Maintenance and/or Repair —Continued

37. Characteristics of polyphase a. c. generators and inverters in relation to load balance and

power factor 1 1 __

38. Principle of operation of magnetic amplifiers. _ 1 1

39. Method of operation, characteristics and functions of under-frequency, overvoltage, and other protective devices used in a. c. distribution systems 1 1 „

40. Method of operation of generator and inverter voltage and speed controls employing electronic

and magnetic amplifiers 1 1

41. Purpose and application of instrument field and instrument special test equipment provided for squadron level maintenance 1 „ 1

42. Application of band pass filters in aircraft instrument amplifiers 1 1

43. Methods of coupling used in electronic amplifiers: transformer, resistive, capacitive, impedance, and direct 1 1

44. Types, application, phasing, and connecting

of transformers in polyphase systems 1 1 C

45. Principles of signal phasing and follow-up as applied in electric automatic pilot servo systems and searchlight control systems 1 1 C

46. Factors causing and methods of locating, suppressing, and eliminating man-made radio

noise interference.— C C C

47. Calculations for power factor correction and

phase balancing in a. c. power systems C C

203 Administrative and/or Clerical

1. Types of information found in NavAer Publications Index 3 3 3

2. Types of information contained in aeronautical technical publications, including operating handbooks, erection and maintenance manuals, allowance lists, parts, catalogs, handbook of maintenance instructions, aviation circular letters, Technical Notes, Technical Orders, Aircraft Service Changes, and bulletins pertinent to instrument and/or electrical systems. . 2 2 2

Qualifications for Advancement in Rating

Applicable Rates

AE A EM AB

203 Administrative and/or Clerical —Continued

3. General content and use of ASO Catalog 2 2 2

4. Maintenance records, logs, and reports required
of squadron maintenance department C C C
5* Applications of military specifications in the installation and inspection of electrical
equipment and wiring in naval aircraft. C C C
6. Use of applicable allowance lists, parts catalogs, and forms in the requisitioning and
accounting for aviation electrical materials C C C

Accelerometer, causes and remedies for troubles with, 245
Adjustable-pitch propellers, 89
Aeroquip fittings, 223
proof test after assembly, 224
Aeroquip tool kit, 224
Aileron control level flight, 168 right wing low, 169
Aileron control of bank-and-climb control unit, 166
Air filter unit, 177
Air pressure measuring instruments, 57
Air, removing from hydraulic system, 179
Aircraft instruments
gyro operation in, 146-148 substitution of, 229
Aircraft plumbing, 221-226
Air-driven gyros, 148
Airplane compass, installing, 119-120
Airspeed indicator (s), 58, 61
causes and remedies for trouble with, 232-233
dial of 430-knot, 63
errors in, 63
mechanism of, 62
reading from, 61, 63 Airspeed tube, sectional diagram,
59
Altimeter (s)
causes and remedies for troubles with, 232-233
checking zero-setting of, 66
mechanism of, 64
standard models, 65
zero-setting error, 66 Altimeter checkup, 67
Altimeter readings, making corrections of, 73
Altimeter reference markers, setting, 67
Altitude above sea, measuring, 64 Altitude controller, barometric, 188
Amplifier, multiple channel, 185 Aneroid barometer, 9 Aneroid, uses, 10 Artificial
horizons, 156 Astro-compass, 204-206
installing, 206 Astrograph
illustration, 207
installing, 208
projection lamps supplied with, 207-208
purpose of, 206 Atmosphere
factors affecting, 6-7
parts of, 6 Atmospheric pressure, 6, 43 Automatic pilot(s), 163

directional gyro control unit of, 172

electric, 180-187

G-3 electric, 187

P-l, 180

type S-3, 163 Automatic pilot oil pressure valve,
179

Automatic pilot preflight check, 187

Automatic pilot servo units, 175 Autosyn fuel flowmeters, 110-112 Autosyn instruments,
functions of, 22

Autosyn operation, basic principle of, 25

Autosyn power supply, 28 Autosyn system, explanation of, 23

Autosyn transmitter, cutaway view of, 26

Bank indicator, 148 Bank-and-climb indicators, 186 Barometer
aneroid, 0

mercurial, 7-9 Barometric altitude controller,
188

Barometric scale reading, 66

Barometric scale, standard range for, 65

Beaded type joint, 222

Bi-metal outside air thermometers, 74

Bi-metal thermometer, 73-75

Bourdon tube, 14

Bourdon type pressure gages, 14, 33

Boyle's law, 11 Bubble sextants, 209 Bulb assemblies, resistance thermometer, 80

Cabin altimeter, causes and remedies for troubles with, 246-247

Capacitor type fuel quantity sage, 112

Capillary thermometers, 76 Chambers, climb indicator, 68 Class 88 aircraft instruments,
repair of, 252 Clftnb indicator (s), 67

chambers in, 68

operation of, 68, 69 Coast test, 155 Code markings, hose, 224 Compass (es)

compensating the, 121-126

direct-reading, 117

earth inductor, 129

flux gate, 129

Compass (es)—Continued

instrument panel, 119

pilot's, 118

radio, 134^136

reading the, 119

remote-indicating magnetic, 126

replacing, 120

swinging, 121

top-reading, 118 Compensators, types, 121 Cross-country flights, setting the
altimeter for, 66 Cushioning material limits, 254r-
256

D-C selsyn fuel gages, 101

De-icing equipment, 13

De-icing pressure gages, 41

Dial markings, instrument, 226

Dial paint, 226

Dial-change indicator, 106

Diaphragm capsule, 62

Direct lift tank unit, 108

Direct-reading magnetic compasses, 117

Directional gyro, 153-155 installing, 154 resetting, 154

Directional gyro card, 163

Directional gyro dial, 155

Drift

measuring over land, 201

measuring over water, 201 Drift angle, 202 Drift sight(s), 199-204

installing, 202-204

Mark 2B, 199-200

Mark 2C, 201

Mark 6, 202-204 Drift test, 155 Dual autosyn indicator

schematic wiring of, 24

transmitters used with, 27 Dual indicators, 38 Dual tachometers, 97-98

Earth inductor compass, 129 Electric automatic pilot, G-3, 187 Electric automatic pilots, 180-187 Electric-driven gyro(s), 148

installing, 152 Electric-driven gyro type indicator, 159

Electrical resistance thermometers, 77-81

Electrical tachometers, 89-93

Electrical units, G-3 automatic pilot, 187

Engine gage unit, 33

Engine-driven vacuum pump, 148

Exposphere, 6

Fast-erecting gyro horizon . indicator

causes and remedies for troubles with, 237-239

gimbal assembly of, 159 Fittings

aeroquip, 223

weatherhead, 223 Flap position indicator, causes

and remedies for troubles with,

245-246 Flared type joint, 221 Flight indicators, 150 Flight instruments, 1 Fluid pressure, 12-14 Fluid pressure systems, 13 Fluid tubing, making bends for, 223

Flux gate compass, 129

Force, definition, 5

Force vs. pressure, 5

Four-engine synchroscope, 97

Free gyro, action of, 144

Fuel and oil pressure indicator, causes and remedies for troubles with, 242

Fuel gages, 101 Ih-C selsyn, 101

Fuel flow transmitter, 111

Fuel pressure gage, 33

Fuel pressure system, 40 Fuel quantity gage system, adjusting, 114

Fuel quantity indicator, causes and remedies for troubles with, 244-245

G-2 compass, causes and remedies for troubles with, 240-241

G-2 compass system, 188

G-3 automatic pilot

barometric altitude controller, 188

location of units, 194

power supply to, 194

stick-type controller, 189 G-3 automatic pilot components,

classification of, 188 G-3 automatic pilot operating

limits, 189-190 G-3 automatic pilot servo drives,

193

G-3 electric automatic pilot, 187 Gage(s) fuel, 101

fuel pressure, 33

liquidometer, 105

manifold pressure, 46

oil pressure, 34

oil temperature, 33

pressure, 33

suction, 43, 44 Gage unit, engine, 34 Generator units

screw type, 92

stud type, 92

tachometer, 89 Generators, mounting, 92

tachometer, 89 Ground tests, instrument, 178 Gyro(s)

air-driven, 148

directional, 153-155

electric-driven, 148

free, action of, 144

274033°—54 19 281

Gyro control units, 166

vertical, 190-191 Gyro flux gate compass, 129

master indicator, 130

parts, 133 Gyro horizon indicator, 44, 156 Gyro-spinning vacuum, 171 Gyroscoi>es, 139-159

elements of, 140 Gyroscopic inertion, 141 Gyroscopic instrumentation, 158 Gyroscopic precession, 141 Gyroscopic principles, 139

Helicopter tachometer with synchroscope, 98 Horizon indicator, dial of, 156 Horizon indicator mechanism, 158 Hose

causes of leaks in, 225 determining location of leaks, 225

inspecting, 225 storing, 225 Hydraulic pressure gages, 45 Hydraulic pump pressure gage, causes and remedies for trouble with, 248 Hydraulic system, removing air from, 179

Indicator

electrical resistance thermometer, 78

magnesyn, 37

oil pressure, 35

tachometer, 89 Inspection form, instrument, 256-257

Instrument dial markings, 226 Instrument flying, 43 Instrument ground tests, 178 Instrument inspection form, 256-257

Instrument landings, 66 Instrument-panel compasses, 119

Instruments aircraft, introduction to, 1 autosyn, 22

cartons for packaging, 253 cleaning, 253

cushioning when packaging, 254 flight, 1 navigation, 1 power plant, 1 precision, 2

shipping and handling, 252 Ionosphere, 6

Joint, beaded type, 222 Joints, types, 221

Landing gear, causes and remedies for troubles, with, 245-246 Leaks, locating in hose, 225 Lighting, red indirect instrument, 227

Liquidometer gages, 105

Lubber's line, 119

Lubricants, approved for aircraft

instruments, 225 Lubrication, aircraft instruments, 226

Magnesyn compass connections,

diagram of, 127 Magnesyn manifold pressure

gage, 50

Magnesyn oil pressure system, 37-39

inspection and maintenance of, 39

Magnesyn system, 29

dual, schematic of, 38

electrical wiring schematic, 30 Magnesyn test transmitter, 250 Magnesyn torque pressure system,

52

Magnesyn transmitter, cutaway

view of, 127 Magnesyn water pressure system,

51-52

Magnetic compasses, 121 Manifold pressure gage(s), 46

readings on, 48

schematic of, 47 Mark 2B pelorus drift sight parts,

199-200 Mark 2C drift sight, 201 Mark 5 sextant, 213-215

illustration, 214 Mark 6-drift sight, 202-204 Median type sextant, 215-217 Mercurial barometer, 7-9 Motors, autosyn, 25 Mounting generators, 92

Navigation instruments, 1, 199 Navigator's top-reading compass, 120

Oil pressure, checking, 179 Oil pressure gages, 34

cold weather maintenance, 36 Oil pressure instruments, 34 Oil temperature gage, 33 Oil temperature thermometers, 78

P-l automatic pilot, 180 components, 181 operating limits, 183 stick type controller, 184 P-l automatic pilot servos, 185 Pascal's principle, 12 Pendulum control unit, 192 Periscopic textant, 210-213

installing, 212-213 Periscopic sextant mount, 210 Personnel, instrument maintenance, 2 Pilots, automatic, 163 Pilot's compass, 118 Pilot's control station, 191 Pilot's instrument-panel compass, 120

Pitot-static tube, 57-450 Plumbing, aircraft, 221-225 Portable field test set, 250

Position indicator, 17 Position indicator unit, installing, 22

Potentiometer, tank unit, 110 Power, adapter, 193 Power plant instruments, 1 Precision instruments, 2 Preflight checks, 178

automatic pilot, 187 Pressure

atmospheric, 6, 43

barometric, 7

definition, 5

fluid, 12-14

measuring manifold, 47 system for indicating, 6 Pressure differences, measuring, 61

Pressure gage(s), 13, 33

de-icing, 41

hydraulic, 45

magnesyn manifold, 50 Pressure measuring instruments, 57

Pressure torqueineter, 52-54

Pressure-sensitive aneroid, 9

Propellers adjustable-pitch, 89 constant-speed, 89

Protective treatment for aircraft instruments, 251-252

Radio compass, general functioning of, 134-136

Range markers, 227-228

Rate and pendulum control unit, 192

Rate-of-climb indicator(s), 10, 58, 67

causes and remedies for troubles with, 232-235

diagram of, 11 Rate-of-climb indicator dial, 68 Rate-of-climb pointer, 70 Red indirect instrument lighting, 227

Remote indicating magnetic compasses, 126 Rotor, directional gyro, 153 Rotor righting mechanism, 159

S-3 automatic pilot installing mounting unit, 175 view of parts, 164

S-3 hydraulic automatic pilot, purpose, 163

Scale correction cards, 67

Screw-type generator, 92

Selsyn d-c system, schematic diagram of, 18

Selsyn pickoffs, 190

Selsyn three-wire system, 19

Selsyn two-wire system, 20-21

Servo disconnect, 186

Servo drives, G-3 automatic pilot, 193

Servo oil system, 174 Servo speed control valves, 175 Servos, P-l automatic pilot, 185 Sextant(s), 209-217

bubble, 209

Mark 5, 213-215

median type, 215-217

periscopic, 210-213

types, 209 Sighting tube assembly, 199 Stand-by compass, causes and remedies for troubles with, 236-237

Static tube, pilot-static system, 58 Steel tubing, 223 Stick-type controller, 184,189,191

Stratosphere, 6 Stud-type generator, 92 Suction gages, 43 Suction gage readings, 55

Synchroscope, four-engine, 96 Synchroscope dial, 95 Synchroscopes, 94-98

Tachometer(s), 89-97 causes and remedies for troubles with, 242-243

Tachometer (s) —Continued dual, 97-98 electrical, 89-93

Tachometer indicator dial, 91

Tachometer indicator mechanism diagram, 90

Tachometer indicators, zero setting attachment on, 93

Tail pipe temperature indicator, causes and remedies for troubles with, 243

Tank unit potentiometer, 109-110

Test equipment, 231

Thermocouple thermometer (s), 81-86

internal wiring of, 85

Thermocouple thermometer indicator, 82

Thermocouple thermometer installation, wiring diagram for, 84

Thermometer (s) f 73

bi-nietal, 73-75

electrical resistance, 77-81

outside air, 73

thermocouple, 81-86

vapor pressure, 76-77 Thermometer bulb assemblies, 80 Top-reading compass, 118

Toroid winding, 30 Torque pressure indicator, 54 Transmitter(s), autosyn, mount-ing of, 29

d-c selsyn, diagram of, 103

fuel flow, 110

magnesyn, 37

magnesyn pressure, 51

three-wire system, 19

two-wire system, 19-20

water pressure, 52 Transmitter unit, autosyn, 24 Troposphere, 6 Tubing bending, 222 •

fitting connections for use on, 222

Tubing—Continued

fluid, bend radii for, 223 rijrid, joints used with, 221

Turn and.bank indicator, 148 causes and remedies for troubles with, 235-236

Universal screw-type compensator, 121

Vaccum, checking in automatic pilot, 178

Vacuum pump method of instrument operation, 148

Vacuum relief valve, 176

Vapor pressure thermometers, 76^-77

Vertical gjro control unit, 190

Volt-ammeter, causes and remedies for troubles with, 248-249

Water injection pressure indicator, causes and remedies for troubles with, 241

Water injection pressure transmitter, causes and remedies for troubles with, 241-242

Water pressure gage, 51

Weatherhead fittings, 223
Wheat stone bridge, 79
Wheatstone bridge diagram, 79
WiHng diagram, synchroscope and tachometer, 95
Zero adjusting knob, 70 Zero-setting error, altimeter, 66